NANOTECHNOLOGY: SOCIETAL IMPLICATIONS I

About the cover

Protein-templated assembly image, courtesy of Andrew McMillan, NASA Ames Research Center (ARC). The computer-generated central image models heat shock proteins that have self-assembled into a double ring structure, 17 nanometers in diameter, called a chaperonin. The researchers can tailor both the chemical functionality of the chaperonin—in this case, the proteins have been genetically modified to bind to nanoscale gold particles—and its structural features. They can coax groups of chaperonins into a variety of 1-, 2-, or 3-dimensional structures, which serve as templates for creating ordered nanoparticle arrays. NASA researchers are exploring the use of protein-templated arrays as sensors and electronic devices.

This is Volume I of a two-volume set resulting from a workshop held under the auspices of the U.S. National Science Foundation and the Nanoscale Science, Engineering, and Technology (NSET) Subcommittee of the U.S. National Science and Technology Council on Dec. 3-5, 2003. The primary purpose of the workshop was to examine trends and opportunities in nanoscience and nanotechnology toward maximizing benefit to humanity, and also potential risks in nanotechnology development. Volume I is a summary of the findings of and discussions at the workshop. The companion Volume II contains essays contributed by the workshop participants.

National Science Foundation

NANOTECHNOLOGY: SOCIETAL IMPLICATIONS I

Maximizing Benefits for Humanity

Edited by

Mihail C. Roco

and

William Sims Bainbridge

National Science Foundation,[*]
Arlington, VA, U.S.A.

Sponsored by

U.S. National Science Foundation and
U.S. National Science and Technology Council
Committee on Technology
Subcommittee on Nanoscale Science, Engineering, and Technology[*]

[*]Any opinions, findings, and conclusions or recommendations expressed here are those of the authors and do not necessarily reflect the views of National Science Foundation or the National Science and Technology Council.

 Springer

A C.I.P. Catalogue record for this book is available from the Library of Congress.

ISBN-10 1-4020-4658-8 (HB)
ISBN-13 978-1-4020-4658-2 (HB)
ISBN-10 1-4020-5432-7 (e-book)
ISBN-13 978-1-4020-5432-7 (e-book)

Published by Springer,
P.O. Box 17, 3300 AA Dordrecht, The Netherlands.

www.springer.com

Printed on acid-free paper

TABLE OF CONTENTS

Appendices

Index

ACKNOWLEDGEMENTS

The sponsors wish to thank all contributors to the report and participants at the December 3-5, 2003 workshop held in Arlington, VA (see Appendix B), particularly the members of the workshop organizing committee:

Davis Baird, University of South Carolina

Vivian Weil, Illinois Institute of Technology

Rachelle Hollander, National Science Foundation

and the staff of the National Nanotechnology Coordination Office (NNCO).

The presentations and discussions at that workshop provided the foundation for this report.

Special credit is due to Cate Alexander and Tom Bartolucci of the NNCO for their editing and technical support, respectively, and to copyeditors Paul D. Lagasse, Paula Whitacre and Joanne Lozar Glenn. Thanks are also due to Geoff Holdridge, Stephen Gould, Philip Lippel, Sam Gill, and other staff members from NNCO and WTEC, Inc. who helped to organize the workshop and contributed to the production of this report.

The work by M. C. Roco and W. S. Bainbridge was performed while serving at the National Science Foundation. Any opinion, findings, and conclusions or recommendations expressed here are those of the authors and do not necessarily reflect the views of the National Science Foundation.

Finally, thanks to all the members of the Subcommittee on Nanoscale Science, Engineering, and Technology, who, through the National Nanotechnology Coordination Office, co-sponsored the workshop with the National Science Foundation, and who reviewed the draft report prior to publication.

This document was sponsored by the member agencies of the Nanoscale Science, Engineering, and Technology (NSET) Subcommittee, through the National Nanotechnology Coordination Office, with particular guidance and participation by staff members from the National Science Foundation. Any opinions, findings, and conclusions or recommendations expressed in this material are those of the authors and do not necessarily reflect the views of the United States Government or the authors' parent institutions.

PREFACE

This report on societal implications of nanoscience and nanotechnology is one of a series of reports resulting from topical workshops convened in 2003-2004 by the Nanoscale Science, Engineering, and Technology (NSET) Subcommittee. The workshops were part of the NSET Subcommittee's long-range planning effort for the multi-agency National Nanotechnology Initiative (NNI), an effort that is informed by broad community input, in part received through these workshops. The NNI seeks to accelerate the research, development, and deployment of nanotechnology to address national needs, enhance our Nation's economy, and improve the quality of life in the United States and around the world, through coordination of activities and programs across the Federal Government.

At each of the topical workshops, nanotechnology experts from industry, academia and government were asked to develop broad, long-term (10 years or longer), visionary goals and to identify scientific and technological barriers that once overcome will enable advances toward those goals. This particular workshop also had the broader goal of identifying and assessing the potential societal impacts of nanotechnology and suggesting ways that related public policy issues might be addressed. The reports resulting from this series of workshops inform the respective professional communities, as well as various organizations that have responsibilities for coordinating, implementing, and guiding the NNI. The reports also provide direction to researchers and program managers in specific areas of nanotechnology R&D regarding long-term goals and hard problems.

This workshop was convened to solicit input and opinions on likely impacts of current and future advances in nanoscience and nanotechnology on the economy, quality of life, national security, education, public policy, and society at large. It also was intended to provide input to the NNI with regard to its research program aimed at understanding and addressing such impacts. The National Science Foundation conducted a similar workshop in September 2000, the report from which was published under the title *Societal Implications of Nanoscience and Nanotechnology*. This workshop was held in December of 2003 to update and extend the findings of the previous workshop.

It should be noted that many people believe that predicting societal impacts is a speculative endeavor for an emerging technology such as nanotechnology. However, past experiences with technological change provide social scientists with insight into potential areas of impact. Theory and modeling based on these experiences are also helpful tools for identifying areas that might be affected by technological changes. It is widely agreed that to the extent we are able to anticipate change, the more prepared we are as a society when such changes arrive.

Given the myriad applications of nanotechnology that are being explored currently, there is a wide range of opinions regarding possible future impacts and social scenarios. This report attempts to reflect the range of opinions expressed at the workshop on the sometimes controversial issues associated with assessing and addressing societal implications of nanotechnology. The reader is encouraged to review these and form his or her own opinions.

The content of this report in no way reflects the considerations of the Federal Government or its constituent agencies, nor any consensus or policy of the Government.

This workshop was originally organized into plenary and breakout sessions focused on areas of potential societal impacts of nanotechnology (e.g., economic, social, ethical, etc.). This report begins with an executive summary followed by an overview of the workshop (Chapter 1) prepared by the report editors, Mihail Roco and William Bainbridge. Introductory and summary comments from the plenary presentations are included in Chapter 2. The results of the 10 breakout sessions are summarized in the "theme reports" in Chapter 3. After the workshop, participants were encouraged to submit written inputs, which the editors have organized into a separate volume, sponsored by the National Science Foundation and published by Springer Science and Business Media.

On behalf of the NSET Subcommittee, we wish to thank all of the workshop organizers, speakers, session chairs, and participants for their contributions to an outstanding workshop and what we believe will be a valuable report. Special thanks are due to William Bainbridge, who edited all of the report chapters in consultation with the various contributors, and to Catherine Alexander in the National Nanotechnology Coordination Office (NNCO), who provided editorial assistance throughout the preparation of the report.

Mihail C. Roco
Co-Chair
Nanoscale Science,
Engineering,
and Technology
Subcommittee

Celia I. Merzbacher
Co-Chair
Nanoscale Science,
Engineering,
and Technology
Subcommittee

E. Clayton Teague
Director
National Nanotechnology
Coordination Office

EXECUTIVE SUMMARY

This report records and synthesizes the opinions of participants at the National Nanotechnology Initiative Societal Implications workshop, held on December 3–5, 2003, in Arlington, Virginia. Sponsored by the National Science Foundation and the Nanoscale Science, Engineering, and Technology (NSET) Subcommittee, the workshop was organized to help the NSET identify trends and opportunities in nanoscience and nanotechnology toward maximizing benefit to humanity, while addressing potential risks in technology development.

Contributing to workshop discussions were experts from industry, government, and academia, representing a wide range of disciplines, including the social sciences, economics, philosophy, physical and biological sciences, and engineering. Topics ranged from economic implications, ethics, public policy, public interaction, education, and workforce development to quality of life, national security and related issues. While environmental and health impacts were an important part of workshop discussions, more technical aspects, in particular, current research and future research recommendations were not addressed directly. Rather, these topics have been the subject of other NNI-sponsored workshops, namely: Nanoscale Processes for Environmental Improvement, Arlington, VA, May 8–10, 2003, and Nanobiotechnology, Arlington, VA, October 9–11, 2003.

Because the term *nanotechnology* covers a very wide range of actual approaches and applications, there cannot be only one type of societal implication or only one policy to promote or to regulate nanotechnology. Science and engineering at the nanoscale have already contributed significantly to many industries, providing new catalysts, coatings, paints, and rubber and tire products, as well as new products and processes in the microprocessor manufacturing, heavy equipment manufacturing, and aerospace industries. Thus, the early implications of nanotechnology are mediated through the numerous other technologies where control over nanoscale structures and processes is advantageous. While the chief short-term effects of these developments are continued economic growth and product improvement, industry and government must also take into account their potential impacts on human health, the environment, and society. The potential exists for major breakthroughs in many areas that could greatly benefit human knowledge, welfare, and security.

A consensus of workshop participants was that an adequately trained scientific workforce is essential for creating and transforming the industries that will realize the benefits of nanotechnology. Accordingly, participants recommended that the United States strive to educate and train sufficient numbers of scientists and engineers. K-16 math and science education must be strengthened, for example, through new curricula and educational materials. Research and educational efforts will require the development of an infrastructure that includes nanoscience laboratories, simulated virtual laboratories, and shared social science information systems that can be used to assess potential societal impacts.

Although there was disagreement over our ability to predict either future advances in nanotechnology or their societal implications, workshop participants generally agreed that the government should fund research to identify potential implications to the extent that such can be determined. Furthermore, the government should attempt to facilitate beneficial impacts and to mitigate negative impacts where they might be expected to emerge. Discussion and debate also focused on how to structure our institutions so that knowledgeable representatives of the public could be involved in decisions regarding technology development.

Participants working in breakout sessions formulated specific analyses and recommendations in 10 thematic areas:

1. Productivity and Equity

2. Future Economic Scenarios

3. The Quality of Life

4. Future Social Scenarios

5. Converging Technologies

6. National Security and Space Exploration

7. Ethics, Governance, Risk, and Uncertainty

8. Public Policy, Legal, and International Aspects

9. Interaction with the Public

10. Education and Human Resource Development

In the social-science field, foundations for reliable social-science understanding and policy guidance have been established in all 10 of the thematic areas, but research is still in the very earliest stages. For example, we do not know the extent to which nanotechnology's implications will be felt through convergence with other technologies—both emerging and traditional—or will be distinctive to nanotechnology itself. We also do not know how broadly the benefits will be enjoyed, either in the short term or long term. Nor do we know if and how cumulative effects could lead to rapid changes.

Participants encouraged additional research on the processes of innovation and diffusion of nanotechnology development. Recommended almost unanimously was physical science research on the possible risks from exposure to nanoparticles and other nanostructures, whether to individual health or the natural environment.

Other research could examine how social and economic forces affect distribution of benefits and risks, across social classes and societies worldwide.

The 10 panels of experts developed a large number of specific recommendations that are included in this report. The following nine are representative:

- Scientifically reliable and publicly respected organizations should clearly articulate the diversity of methods and principles of nanotechnology, as well as the near-term and the long-term benefits and uncertainties of nanotechnology, in order to solidify public trust and empower people to participate in public-policy discussions and decisions about nanotechnology.

- Research should

  - be based primarily on peer-reviewed investigator-initiated proposals and should not be driven by a few specific top-down priorities, but rather should reflect the breadth of potential applications of nanoscale science and engineering

  - be supported to develop various models of public involvement and interaction, to establish successful methods for educating, communicating, and engaging diverse publics about nanotechnology

  - incorporate ongoing engagement of the public in deliberations on nanotechnology to assure two-way interchange between nanoscientists and engineers and the public

- The government should

  - review research aimed at understanding the human health and environmental consequences of nanomaterials and adjust funding as necessary to address areas where more information is needed

  - review the adequacy of the current regulatory environment for nanomaterials, given the existence of size-dependent properties

  - develop a communication strategy to keep the public informed of representative and fundamental developments of the new technology

- The government and the private sector should assess potential implications and scenarios of new technology development and communicate those assessments to policy makers and the public for potential response. An example would be assessments of industries and jobs that could become obsolete. Such assessments could facilitate transitions through education, retooling, and similar efforts.

- A careful and rigorous analysis of the adequacy of current NNI funding levels and of future investment priorities is necessary to optimize societal benefit.

- Nanotechnology educational initiatives should aim to support graduate and postdoctoral students through cross-disciplinary training and experience;

models for collaboration among physical and biological scientists and engineers, social scientists, and humanists across disciplines; and integration of social science and technical research. Nanotechnology education should be provided beginning with K-12 programs.

- To meet current and short-term labor needs, the government should support the implementation of training programs to equip underutilized scientists and engineers with nanotechnology-related skills.

- Increased capabilities and funding should be developed for conducting science and technology studies in educational contexts, in industrial contexts, and among the public. Workforce development should be undertaken across the full spectrum of job roles, not just among research scientists.

- A programmatic approach is needed to increase synergy in nanotechnology development by creating partnerships earlier in R&D processes between industry, academia, national laboratories, and funding agencies, as well as corresponding international organizations. Multi-functional clusters or partnership coalitions with greater flexibility to adapt should be created that bring together those involved in researching and developing nanotechnology with those working in other fields, such as biotechnology and information technology.

One view expressed repeatedly during the workshop is that the National Nanotechnology Initiative can play an important role in coordinating research and development in nanotechnology while addressing public hopes and fears. Workshop participants support NNI efforts to build capacity for public dialogue and to deal with genuine risks in an expeditious, open, and honest manner.

1. OVERVIEW

Mihail C. Roco and William Sims Bainbridge

INTRODUCTION

Nanotechnology development offers the promise of advances within and connections across many disciplines—from physics, to chemistry, biology, materials science and engineering [1]. The new technology will provide a broad technological platform for industry, medicine, and the overall economy. Much like information technologies, nanotechnology is expected to be embodied in many products. "The product of nanotechnology is not itself a final product," as keynote speaker George W. Whitesides noted, "but it goes into something—for example, a computer—that becomes a product."

According to industry experts at the workshop, within 10 years nanotechnology could be used in nearly half of all new products, from handheld computer devices to cancer and other disease treatments; renewable energy sources; lightweight multifunctional components in cars and airplanes; agents for environmental remediation; and water filters that remove viruses, contaminants, and salt for entire cities. Such potential strides explain why nanotechnology is viewed as key to future economic growth and why technologically advanced countries are earnestly pursuing its development across the globe.

While research points clearly to many potential applications, societal changes that could result from this versatile technology are less well understood. The purpose of this workshop was to identify future trends and research opportunities related to the societal changes resulting from nanotechnology, both positive and negative. Drawing from the workshop, this report surveys the current state of knowledge about societal implications and explores potential developments over the next 10 years and beyond related to economics and education, as well as social, ethical, and legal issues. The report identifies the areas of research, education, and infrastructure development that would be most valuable for society, and suggests methods of investigation that are most appropriate for both research and program evaluation. The dozens of scientists, engineers, and policy leaders who contributed to the Societal Implications workshop, recommend actions and anticipatory measures that they believe are needed to take prompt but responsible advantage of the new technology. Their recommendations are included in this report.

The early involvement of social scientists, ethicists, humanists, and others in defining the attributes of responsible development of nanotechnology, is both necessary and a relatively new phenomenon. Rather than being confined to the role of commentators who observe a technological development as it proceeds, social

1

M.C. Roco and W.S. Bainbridge (eds.),
Nanotechnology: Societal Implications — Maximizing Benefits for Humanity, 1–13.

ANTICIPATED ADVANCEMENTS

The list of potential advancements offered by nanotechnology is quite varied. The following illustrates the wide range of areas where nanotechnology innovation is expected:

- Improvements to computing, sensing, communications, data storage, and display capacities. Among anticipated developments are automatic extraction of information from raw data, artificial intelligence, and virtual reality.

- Substantial contributions toward energy independence for the United States and other energy-consuming nations, both from ecologically sound production of energy and from a reduction in the demand for energy caused by a host of efficiencies facilitated by nanotechnology.

- Advanced, high-performance robotics relying on nanoscale components, leading to the creation of new medical devices, smart unmanned platforms for deep space exploration, and combat vehicles with minimal risk to human crews.

- Composite materials with a high strength-to-weight ratio, facilitating very-high-performance space launchers and fighter aircraft that achieve low life-cycle costs.

- New biomedical solutions for chronic diseases, new drugs and targeted drug delivery, and visualization of key biological processes within the human body.

- Affordable nanoscale miniaturized medical diagnostic and treatment devices that contribute to increased longevity with greater comfort, activity and productivity, easing medical hardships, pain, suffering, and disabilities.

- Protection for persons in hazardous environments, such as soldiers on the battlefield, including clothing that incorporates nanoscale devices to constantly monitor physiological vital signs, warn of exposure to harmful chemicals, adjust for environmental stresses, provide camouflage that matches changing background and lighting conditions, and even provide first-aid casualty response.

scientists are being asked to provide valuable input and perspectives that will guide the relevant scientific communities as they attempt to ensure that their discoveries provide maximum benefit to humanity while limiting potential risks.

1. Overview

This report addresses a central question posed by technological development: how does society design and employ advances for a better tomorrow, while preserving what is highly valued by citizens today? Another of the workshop's keynote speakers, former National Science Foundation Director Rita Colwell, spoke directly to this when she said, "We need to anticipate and guide change in order to design the future of our choice, not just one of our making."

This issue—the human capability to advance technology vis a vis the human capability to anticipate and plan for outcomes—is relevant to all technological innovation, including nanotechnology. Because technology and society are complex, interacting systems, it is difficult to predict the long-term consequences of particular innovations. Yet some degree of predictability is necessary in order to plan for change. While innovation does not "just happen" absent identifiable processes, theory and models for the rational management of technological innovation are lacking and must be developed. As technological change is expected to accelerate, theoretical and modeling work must keep pace with the development and deployment of new technologies. Scientists must ensure that they provide policy makers with the scientific information necessary to make their decisions. Policy makers must in turn ensure that the appropriate governance mechanisms for anticipation and correction are in place to address unexpected consequences.

There is a broad consensus that rational management of the innovation process, including nanotechnology innovation, must involve a variety of stakeholders beyond the scientific community, including representatives of the general public. The wide range of interests in society must provide value-based inputs that can be used to balance economic development needs with those of human health, the environment, and, more broadly, the quality of life. Feedback—from a well-informed public and from international collaboration—has become essential for progress. More interactions between scientists, economists, and the public are needed to identify and reach the robust balance between benefits and risks that apply to innovative technologies, including nanotechnology.

The first significant effort to understand the societal dimensions of nanoscience and nanotechnology was an earlier workshop held on September 28-29, 2000, at the request of the Subcommittee on Nanoscale Science, Engineering, and Technology (NSET) and organized by the National Science Foundation [2]. During the intervening three years, nanoscience progressed rapidly, and related discoveries were applied to products sooner than many had expected. As a result, a consensus has emerged in the scientific and non-technical communities that, once an obscure field, nanotechnology now makes and will continue to make significant contributions to technological development.

3

This realization has led to several other reports on the societal dimensions of nanotechnology, including one that documents a European Union-United States meeting on the subject in 2002 [3], and another on the International Dialogue on Responsible Nanotechnology Research and Development sponsored by the NSF with the Meridian Institute in 2004 [4]. Other reports on societal dimensions have been issued by the ETC Group [5], Greenpeace [6], the European Community [7], the Swiss Re company [8], and VDI of Germany [9]. In addition, several working groups have been established, including the Nanotechnology Environmental and Health Implications working group within the NSET, the Consultative Boards for Advancing Nanotechnology (CBANs) convened jointly by NSET and various industry sectors in the United States, the International Council on Nanotechnology (ICoN) at Rice University, a working group created by the Royal Society and Royal Academy of Engineers in the UK, and a similar organization in Switzerland.

This report comprises the Executive Summary, Overview, Introductory and Summary Comments of keynote speakers, and Workshop Breakout Session Reports on 10 transformative themes. In addition, a separately published report sponsored by NSF contains about 50 individual contributions of workshop participants, which further examine important aspects of nanotechnology development.

Two other NNI-sponsored reports complement this report: one on environmental aspects [10], and another on nanobiology and nanomedicine [11].

TEN TRANSFORMATIVE THEMES

Participants in the December 2003 Societal Implications of Nanoscience and Nanotechnology workshop explored and made recommendations in 10 thematic areas: productivity and equity; future economic scenarios; the quality of life; future social scenarios; converging technologies; national security and space exploration; ethics, governance, risk and uncertainty; public policy, legal and international aspects; interaction with the public; and education and human resource development. Each of the 10 panels provided a summary of its deliberations, which are included in Chapter 3.

The first five of these themes are primarily concerned with the benefits that nanotechnology could provide to humanity, and some of the problems it might cause if it is not developed wisely. Three of the themes concern the broad issue of nanotechnology investment and risk governance, including ethical and legal issues, policy-making institutions, risk assessment and management, and appropriate ways to include the general public in the decision process. The final theme, education and human resources, is in many ways the most fundamental. Without enough well trained scientists and engineers and educated citizens, the benefits of nanotechnology might go unrealized. A widespread, accurate awareness of the basic facts of nanoscience will provide the public and policymakers with the knowledge they need to make informed decisions about nanotechnology and its products. Here are brief abstracts of the panel reports:

1. Productivity and Equity

Nanotechnology has entered the mainstream of industry and many companies have shown their confidence in nanotechnology's future by committing substantial resources to its development. In the near term, developments are expected to be gradual changes that will incrementally improve manufacturing costs and product features. Longer-term developments are likely to occur in convergence with other emerging technologies, such as biotechnology and information technology, where nanotechnology will serve as an enabler of a new product or industry category.

The effects of nanotechnology are expected to stimulate productivity in manufacturing in most sectors of the economy that deal with the materials world. Better tools and measures for understanding the social and economic implications of nanotechnology are recommended. Researchers must proceed beyond the use of published, aggregate-level data available in econometric studies in order to get inside the research and development (R&D) processes as they occur. The panel recommended a programmatic approach to increase synergy in nanotechnology development by creating partnerships earlier in the R&D processes between industry, academia, national laboratories, and funding agencies. To the extent possible, government and the private sector should anticipate and mitigate negative impacts resulting from new technologies, including worker displacement and unbalanced distribution of benefits and risks in society.

2. Future Economic Scenarios

In addressing future economic effects, researchers need to take a number of different, but ultimately complementary, viewpoints. The overall macroeconomic viewpoint considers the effects on economic growth, productivity, real wages, and the standard of living. The industrial organization viewpoint focuses on the particular industries that will be most directly affected by nanotechnology and attempts to assess how they will be transformed. For each of these viewpoints, the first step is to frame the "counter-factual" question—what is the effect of the nanotechnology compared to what alternative?

To maximize the benefits from investment in nanoscale science and engineering, scientific research should be broadly funded and based primarily on peer-reviewed investigator-initiated proposals. Research should not be driven by a few specific top-down priorities. Economists can help to maximize development of beneficial applications by contributing research in the following areas as applied to nanotechnology: the transfer of knowledge from academe to industry; the levels of return on nanotechnology investment; the effect of healthier lives on work patterns; the skill biases associated with major nanotechnology applications and their implications for wages and returns to education; the potential impacts and prudence of research exemptions for patents; and, lastly, the most efficient and effective forms of government-industry-academe research cooperation.

3. The Quality of Life

Research on the societal dimensions of nanotechnology should identify the qualities of work, life, and the environment to which citizens give their highest priority, and identify the branches of nanotechnology that are most relevant to them. A means of monitoring for the early signs of negative aspects and risks should be developed, thereby permitting the timely development of contingency plans to handle problems. An issue of great political and ethical significance is the possibility that improved capabilities deriving from nanoscience will cause the gap between the haves and the have-nots to grow. Research should seek to understand the conditions under which this could happen, as well as to identify the factors that can maximize the distribution of nanotechnology's benefits throughout the population. Additionally, research should seek to identify the kinds of institutions that are best able to safeguard the public in areas such as privacy and nanotechnology-related hazards, without inhibiting development of beneficial applications. Scientists, educators, the mass media, and policy makers should clearly distinguish the direct effects of nanotechnology on the quality of life, as distinct from the effects of other technologies (such as genetic engineering) that are sometimes connected to nanotechnology, but are in fact separate phenomena requiring separate treatment.

4. Future Social Scenarios

New technologies do not merely have implications for society. Rather, new technologies interact with society, and their impact is the result of the interaction of multiple technical facts with social factors. At present it is difficult to distinguish between those impacts that are distinctive to nanotechnology and those that arise from the synergistic effects of the convergence among technologies. Nevertheless, effective theoretical and modeling tools can be used to assess potential impacts, thereby providing policymakers with the information they need to evaluate whether our institutions will be prepared to take the best advantage of the positive impacts while reducing the societal or economic costs of any negative ones.

Scenario analysis can help identify issues and hypotheses and is thus a useful tool for theoretical analysis. Multi-agent modeling is akin to scenario analysis, but is carried out through computer simulation. Research is recommended to compare the history of nanotechnology with the history of other technologies having significant positive and negative aspects, such as that of genetically modified organisms, stem cells, and nuclear power, to draw lessons that might be relevant. Research on the processes of innovation, diffusion, and adjustment will also be important.

5. Converging Technologies

The unity of nature at the nanoscale provides the fundamental basis for the unification of science, because many structures essential to life, computation, and communication are based on phenomena that take place at this scale. Thus, much of the impact of nanotechnology will occur through its convergence with other fields, especially biotechnology, information technology, and new technologies based on cognitive science. The power of converging technologies offers the potential for enormous societal benefits including economic growth, job creation, national defense, homeland security, and improvements in a variety of other areas. The same power, combined with the uncertainties of working at the leading edge of a technology, raises significant questions about risk assessment and management. Researchers need better models for risk analysis, characterization, and quantification. The maximization of benefit to humanity will require the development of transforming tools that can be shared across fields, including: new scientific instrumentation, overarching theoretical concepts, methods of interdisciplinary communication, new organizations and business models including multifunctional technology platforms, and fresh techniques for production such as those that bridge the gap between the organic and the inorganic. Technological convergence is the wave of the future, but it cannot properly transform science and technology without the investment of considerable effort to achieve maximum benefits for humanity.

6. National Security and Space Exploration

Nanotechnology provides advantages for almost every aspect of operations required for national security and for space exploration, in terms of the ability to gather, communicate, digest, and act upon information with advanced sensors, and to take requisite action with platforms that will have augmented capacity. In both arenas, there are also pressing needs for stronger, lighter, more durable structural materials and for reliable explosives and propellants that release greater energy. Nanotechnology can provide advanced materials for aircraft, armor for combat or re-entry vehicles, and ships or satellites that are less vulnerable to corrosive environments. Advances in these areas will lead to commercially available spin-off products and services. A well-educated workforce is a key to realizing these developments. Economists and social scientists should develop an accurate estimate of scientific workforce needs, and the government should support the implementation of retraining programs to equip underutilized scientists and engineers with nanotechnology-related skills.

7. Ethics, Governance, Risk, and Uncertainty

Research generally shows that the "news model" of public involvement, in which technical experts and the media impart information to a passive audience, is not a highly effective means of informing the public. Information systems that allow two-way conversation may achieve greater levels of understanding. Government, industry, and academia can create opportunities for conversation between

ANTICIPATED SOCIETAL DEVELOPMENTS

While true solutions to societal problems require enlightened policy, world peace, and equitable distribution of resources, public and governmental interest in nanotechnology R&D is providing a model for responsible technological development. Among the societal changes that were envisioned at the workshop as contributing toward responsible development of nanotechnology are:

- Substantial contributions to economic growth

- Widespread public involvement in discussions and deliberations about technology development

- Advanced and educationally effective online-information resources devoted to ethics and societal dimensions of nanotechnology, integrated not only into conventional K-12 and college courses but also into continuing education for companies, scientists, and engineers

- Interactions between teams developing scientific, engineering, or social projects related to nanotechnology and expert communicators, who communicate across fields, facilitating multidisciplinary collaboration

- Innovations and reforms in many branches of the law, including torts, environmental law, employment and labor law, health and family law, criminal law, constitutional law, international trade law, and antitrust law

- Supported by public understanding and confidence, successful coordination by the National Nanotechnology Initiative of nanoscience research and development to achieve the social, technological, and economic goals driving the development of nanotechnology

nanotechnology specialists and members of the public. A range of projects could develop infrastructures for balanced and inclusive public participation in decision making with many different, innovative models used to assure two-way interchange between nanoengineers or scientists and their publics. Fundamental knowledge about the origins and functions of interest groups is lacking. Research is needed to gain such knowledge, as well as to gain information on how publics evaluate information and on nodes of controversy among informed and uninformed parties, with attention to cross-cultural differences. The panel also notes that society's efforts to shape the future—rather than simply trying to predict it—could prove more fruitful by building institutions that can learn while preserving core values.

8. Public Policy, Legal, and International Aspects

The panel recommends a careful and rigorous analysis of the adequacy of current funding levels for nanoscale science and engineering, as well as safety research, legal aspects, and other issues. Because nanotechnology research has the potential to play a considerable role in mitigating the problems of the developing world, many legal or policy issues will need to be addressed on a global, rather than national, scale. Nanotechnology has implications for legal work, including not only intellectual property and the commercialization and technology transfer of research, but also the wide integration of multiple legal practice areas. As nanotechnologies reach the commercial marketplace, the public should be confident that the government is taking appropriate steps to safeguard the environment and human health, while also enabling new technologies and new industries to flourish. Therefore, government-supported research is necessary to understand the human health and environmental consequences of nanomaterials and to review the adequacy of the current regulatory environment for these materials.

9. Interaction with the Public

Negative public attitudes toward nanotechnology could impede research and development, leaving the benefits of nanotechnology unrealized and its economic potential untapped, or worse, leaving the development of nanotechnology to countries and researchers who are not constrained by regulations and ethical norms held by most scientists worldwide. The National Nanotechnology Initiative can play an important role as an honest broker in coordinating research and development in nanotechnology, dealing with genuine risks in an expeditious, open, and honest manner.

Research is required on the best way to achieve an informed population so that successful approaches for educating, communicating, and engaging diverse publics about nanotechnology can be established. Research is needed on public understanding and attitudes, audience response to various media products, and effective training methods to prepare scientists and engineers to engage in public dialogue about nanotechnology. Readily available educational opportunities would provide publics with a means to become actively engaged in learning about nanotechnology. Building capacity for public dialogue means developing repositories of knowledge accessible to citizens as well as creating physical places, such as museums and science and technology centers, where citizens can meet and learn from each other's perspectives, including the perspectives of those who are pursuing nanotechnological research and those who are engaged in developing nanotechnology-related policy. Social scientists and humanists also must examine how to frame risk communication regarding nanotechnology and how public responses are influenced by the way risks are framed.

10. Education and Human Resource Development

Nanotechnology creates an opportunity to integrate education across physical and biological sciences, technology, the social sciences, and even the humanities. Such integration is emblematic of new ways of thinking about the future and the workforce. Students will be motivated by problems that combine the social and the technical—for example, the potential for new environmental technologies. Nanotechnology studies, especially through the convergence of many fields of science and engineering at the nanoscale, will contribute to fulfilling the mission of liberal education to make students into critical thinkers, capable of participating in intelligent debates about how societies ought to be transformed. The end result will be informed, educated publics emerging from our high schools and colleges, able to shape the direction of nanotechnology in beneficial ways. Research needs to focus on the viability and transferability of strategies that can be integrated across disciplines, including curriculum test beds where students and teachers could work with nanoscientists. Disciplinary education will need to be complemented by training in ways of working in interdisciplinary teams. Worker transition programs should be framed as opportunities to get in on the ground floor of a growing field. The societal dimensions of nanotechnology create an opportunity for introducing new concepts to students from grades K to 16, and for training postdoctoral and other advanced students in areas of technology that will be in demand from society.

KEY RECOMMENDATIONS AND RESEARCH OPPORTUNITIES

Perhaps not surprisingly, one of the overarching recommendations from the panels and individual contributors to this report is that policy makers should ensure the development and deployment of beneficial applications of nanotechnology, while anticipating and avoiding potential negative impacts. Another important recommendation was that entire societal systems should be addressed in both the short and long term, rather than addressing societal implications one-by-one.

Contributors stressed the need to change educational systems significantly so that new technologies, including nanotechnology, will be integral to science curricula for all ages. They stressed the need for an educational system that nurtures the ability to think across disciplines and to communicate well. They also emphasized the need to assess weaknesses in the existing workforce so that retraining and supplemental training of scientists, engineers, and others will contribute to technological development. Because nanoscience and nanotechnology are being developed in all industrialized nations, no country can depend upon foreign students for its scientific human capital; thus the United States must produce an ever-increasing number of domestic scientists and engineers.

Nanotechnology presents both the need and the opportunity for transformation of our educational system. Teaching of the sciences is highly fragmented today, whereas nanotechnology bridges physics, chemistry and biology. Some believe that if we first integrate science teaching around phenomena at the nanoscale, education could likewise be integrated across the physical sciences, technology, the social sciences and even the humanities [12, 13]. The challenges are immense, but the need for a new model of education is widely recognized.

A second research area emphasized by nearly every breakout group at the workshop (and by many individual contributors to this report) was that of ascertaining risks associated with nanotechnology. All technologies present advantages and disadvantages and must be managed; however, the idea of controlling materials at the nanoscale, which is at the same scale at which living systems operate at their most fundamental level, raises concerns for some people. Research is needed to understand the risks, how to mitigate those risks that are real, and how to harness nanotechnology to the service, but not the disservice, of humanity.

Of equal importance to ensuring the development of nanotechnology products that enhance the quality of life, is the need to engage the interested public in discussions about such development. The issues of power and trust are central to public dialogue, and to merely inform the general public about the conclusions of experts is insufficient to shape public policy effectively. Public engagement—and by this we mean engagement with many groups with special concerns and interests—must be an active process of deliberation.

Opinion polls and surveys can determine where the public stands on nanotechnology-related issues, chart the changing perceptions of nanotechnology over time, and compare public understanding of technology issues across different cultures. Methodologies such as focus groups, consensus panels, case studies and interviews can provide policy makers with useful information about public values and concerns about new technology development.

From consideration of the early applications of nanotechnology, in light of the research on other technological developments of the past, economists and other social scientists are able to project some very general trends. With the passage of time, progressively more advanced nanotechnologies can be expected to lower manufacturing costs in some industries, improve productivity in others, and even create entirely new industries. The result will be increased demand for some goods and skills, and reduced demand for others. While the short-term result may mean disruption of some specific corporations and careers, the economists who participated in the Societal Implications of Nanoscience and Nanotechnology workshop were optimistic that the free market system would ensure that capital and labor will shift to new uses, and the disruption will be limited to a transition period in narrow sectors of the economy. On balance, they believe, technological development will improve the standard of living, both through economic growth and through the new capabilities provided by the technology.

It will be necessary to prioritize which of the numerous interesting research topics, potential education programs, and valuable kinds of infrastructure should receive the earliest or greatest investment.

A high priority clearly should be given to physical science research on the possible risks—whether to individual health or the natural environment—from exposure to nanoparticles and other nanomaterials. Research on the processes of innovation, diffusion, and adjustment would certainly be useful, as would a better understanding of the complementary influences of private, venture, university, and government investment in nanotechnology development. Other research should examine how social and economic forces affect distribution of benefits and risks, both across social classes and across societies of the world.

Current laws, regulations, policies, the educational preparation of legal professionals, and judicial infrastructures will be challenged, as well. The adequacy of our institutions, organizations, and laws for achieving the best balance of innovation, security, equity, health, and environmental protection is a question that these editors hope to see evaluated.

REFERENCES

1. M. C. Roco, R. S. Williams, P. Alivisatos, eds., *Nanotechnology Research Directions*, Washington, D.C.: NSTC (1999), also Dordrecht: Springer (formerly Kluwer) (2000).

2. M. C. Roco, W. S. Bainbridge, eds., *Societal Implications of Nanoscience and Nanotechnology*: National Science Foundation Report, Arlington, VA: National Science Foundation (2000), also Dordrecht: Springer (formerly Kluwer) (2001) (available at http://www.wtec.org/loyola/nano/NSET.Societal.Implications/).

3. M. C. Roco, R. Tomellini, eds., *Nanotechnology revolutionary opportunities and societal implications*, Luxembourg: European Communities, Office of the Official Publications of EC (2002).

4. Meridian Institute, *International Dialogue on Responsible Research and Development of Nanotechnology*, Alexandria, VA (2004) (available at http://www.nanoandthepoor.org/Final_Report_Responsible_Nanotech_RD_040812.pdf).

5. ETC, *No small matter II: The case for a global moratorium—size matters!*, ETC Occasional Paper Series **7**, p1–14 (2003).

6. Greenpeace, *Nanotechnology*, Greenpeace (2004) (available at http://www.greenpeace.org.uk).

7. European Commission, *Nanotechnology: A preliminary risk analysis on the basis of a preliminary workshop*. Paper read at a workshop organized by the Health and Consumer Protection Directorate General of the European Commission, March 1–2, 2004, in Brussels, Belgium (available at http://europa.eu.int/comm/health/ph_risk/documents/ev_20040301_en.pdf).

8. Swiss Re, *Nanotechnology: Small matter, many unknowns*, Switzerland: Swiss Re, (2004) (available at www.swissre.com).

9. L. Wolfgang, ed., *Industrial application of nanomaterials—Chances and risks*, Dusseldorf, Germany: VDI Technologiezentrum (2004).

10. NSET, *NNI Grand Challenge Workshop: Nanoscale Processes for Environmental Improvement*, May 8-10, 2003, in Arlington, Virginia.

11. NSET, *NNI Workshop on Nanobiotechnology*, October 9–11, 2003, in Arlington, Virginia (available at http://www.nano.gov/nni_nanobiotechnology_rpt.pdf).

12. E. O. Wilson, *Consilience: The Unity of Knowledge*, New York: Knopf (1998).

13. M. C. Roco, W. S. Bainbridge, *Converging Technologies for Improving Human Performance: Nanotechnology, Biotechnology, Information Technology and Cognitive Science*, Dordrecht: Springer (formerly Kluwer) (2003).

2. INTRODUCTORY AND SUMMARY COMMENTS

WELCOME

Rita R. Colwell, [former] Director, National Science Foundation

Welcome to the National Science Foundation (NSF). The goals of this workshop are lofty. Your charge is to help prepare the Nation for new scientific, engineering, and technological capabilities forged at the scale of individual atoms and molecules. NSF's commitment is to support you in meeting that goal.

Almost 5 percent of our [fiscal year 2004] $5.5 billion budget request is targeted at strengthening the U.S. leadership in nanoscale research and education. Our investment is leveraged through partnerships with other organizations, investments by other agencies, and relationships with the private sector. Most of NSF's investment—indeed, much of the research across the Nation—is focused on the first step in the science and engineering process, developing knowledge and capability.

There's a lot of discussion about specific products expected from nanotechnology, such as environmental sensors or new means of delivering drugs. However, although we know more than we did a few years ago, we can't know the full scope of tools and materials that can be created by the manipulation of nanoscale building blocks. We are just beginning to explore the potential of what we are learning.

The science and engineering community is already developing manufacturing and business expertise. We know that global competition will be fierce, and the United States must be prepared. There is also a simultaneous need to examine the social and ethical aspects of research that could ultimately transform the way we live and work. As we strive to advance nanoscience and engineering, we must do so benignly and equitably. This will require active involvement with the social sciences and with concepts for managing risk.

NSF has begun this process by investing in exploratory grants for the study of societal, ethical, and security implications. This workshop and the earlier one in 2000, sponsored by NSF and the National Science and Technology Council's NSET subcommittee, are intended to help steer the growing nanoscience and engineering community in appropriate directions.

NSF's newest budget priority on human and social dynamics will help facilitate studies in how to manage change invoked by the new science and engineering; whether change occurs in the workplace, the environment, in healthcare, or in areas that have not yet reached the research stage.

15

M.C. Roco and W.S. Bainbridge (eds.),
Nanotechnology: Societal Implications — Maximizing Benefits for Humanity, 15–51.
© 2007 *Springer.*

These issues are not ancillary to the science, engineering, and technology enterprise. Research is about exploring, taking risks, and embracing new knowledge. Our broader mandate includes informing the public and engaging broad segments of society in decisions on how the new capabilities should be applied.

We need to anticipate and guide change in order to design the future of our choice, not just one of our making. We want society to be prepared for, though not necessarily control, the results of far-reaching research. Future generations may well judge our success—and our wisdom—by how well we realize the potential of nanoscience and engineering while avoiding the pitfalls.

In each NSF program or initiative, we embody the need to forge a seamless transition from research results to societal change. We must act quickly as a community to develop the means for measuring and predicting the societal and economic impacts of nanoscale research. Following the pioneering work of Richard Smalley, nanotubes quickly became a recognized commodity. Fine-grained powders manufactured at the atomic level also found a ready market. What are the next outcomes, the next products? The answers are coming swiftly, and we need to be prepared.

As we advance the technology, we must ready the social infrastructure and engage the public in managing change. It's a weighty challenge, but one that scientists and engineers already engaged in far-reaching research can readily embrace. We welcome your ideas and assistance.

THE FUTURE OF NANOTECHNOLOGY †

John H. Marburger, III, Director, Office of Science and Technology Policy

Good afternoon. My instructions are to provide a "visionary presentation focusing on the future of the field" of nanotechnology from the perspective of the Federal Government. Let me say at the outset that "nanotechnology" is not so much a "field" as a word—a neologism—that has been pressed into service to symbolize the status of a very large and important sector of contemporary science. It is possible to find narrower uses of the word, some of which I will mention later, but participants in this workshop are surely aware that nanotechnology refers implicitly to a set of capabilities at the atomic scale that grew steadily throughout the last half of the past century into the basis for a true technology revolution in our society.

I speak of a revolution in technology rather than in science, because the underlying science is not itself revolutionary. Not that the scientific work is complete or unexciting—to the contrary—but we have known for more than a century that all the

---

† This talk was presented by Celia Merzbacher of OSTP so that Dr. Marburger could be present at the signing of the 21ˢᵗ Century Nanotechnology Research and Development Act.

matter of everyday life is made of atoms. And we have known in principle for nearly 50 years how to calculate, with exquisite precision, many of the properties of matter, given certain input information and enough computing power. But not until recently have we actually had the instruments to make atomic-level measurements, and the computing power to exploit that knowledge. Now we have it, or are getting it, and the implications are enormous. Everything being made of atoms, the capability to measure, manipulate, simulate, and visualize at the atomic scale potentially touches every material aspect of our interaction with the world around us. That is why we speak of a revolution—like the industrial revolution—rather than just another step in technological progress.

It is no wonder that developed nations are eager to produce and acquire the technologies that are being spawned by these new atomic-scale capabilities. As far as I can tell, the science plans for every developed nation and the European Union have a strong "nano" focus. The United States has been a world leader in the development of the underlying science infrastructure for the revolution, and the development of nanotechnology is a national priority today. Let me take a few moments to put the National Nanotechnology Initiative (NNI) into perspective.

The revolution implicit in nanotechnology extends to all functional behavior of material that is influenced by nanoscale structure. That includes much of biology, chemistry, and materials science. These subjects have long histories, and therefore many of the issues we are gathered to discuss today are already familiar to us in other guises. This has implications that I will describe shortly. For now, I want to point out that federal funds have supported investigations into nanoscale phenomena in biology, chemistry, and materials science for a long time. Much, but not all, of biotechnology is what I like to call "wet nanotechnology." A growing fraction of Federally funded biomedical research extending over four decades deals with the life processes in humans and other organisms at the molecular and nanoscale levels. This Administration's substantial increase in federal funds for medical research— now consuming nearly half the total federal science and technology budget—is justified in large part by the enormous promise of biotechnology as applied to medicine.

The instrumental and computational infrastructure that provides the new nanoscale capabilities has also been built up with Federal funding over decades, particularly in the Department of Energy multi-program laboratories, but with significant facilities also at university centers funded by the National Institute of Standards and Technology (NIST) and National Science Foundation (NSF). This infrastructure includes electron microscopes, bright X-ray sources, nuclear magnetic resonance devices, mass spectrometers, scanning probe microscopes, and a variety of optical and infrared spectroscopic devices. The infrastructure also includes the inexorably improving power of computation, communication, and data storage capabilities that we lump together under "Information Technology." Today, the development of information technology and its application to nanoscale technologies are national priorities.

In the current fiscal year, the President requested, and Congress is likely to fund, the NNI at approximately $850 million spread among 10 Federal agencies, of which the National Science Foundation manages the largest share. This interagency program is coordinated through the Subcommittee on Nanoscale Science, Engineering, and Technology (NSET), which is the mechanism established for such purposes and managed by the Office of Science and Technology Policy (OSTP). Congress very recently passed a nanotechnology bill (S189) that establishes important new principles for the pursuit of this broad area of science and technology. On this very day, President Bush will sign this bill into law in a special ceremony at the White House.

This audience should be aware that the nanotechnology bill includes a number of provisions related to societal concerns. It requires, for example, that the program ensure "that ethical, legal, environmental, and other appropriate societal concerns, including the potential use of nanotechnology in enhancing human intelligence and in developing artificial intelligence which exceeds human capacity, are considered during the development of nanotechnology ... " [1, section 2.B.10]. The bill also requires

- the establishment of a research program on these issues

- that societal and ethical issues be integrated into all centers established by the program

- that public input and outreach be integrated into the program

A provision to set aside 5 percent of overall program funding to study societal and ethical issues was defeated during markup of the bill in the House Science Committee, but the proposal indicates how seriously Congress takes these issues. The bill charges a Presidential advisory committee with determining and reporting biannually to the President whether social and ethical concerns are being "adequately addressed by the program" [1, section 4.C.7].

The bill further requires two studies by the National Research Council (NRC), one on the technical feasibility of "molecular self-assembly for the manufacture of materials and devices at the molecular scale," [1, section 5.B] and another on the responsible development of nanotechnology. Finally, the bill requires a center focused on societal and ethical issues of nanotechnology.

This is heavy machinery and indicates an extraordinary level of interest in these issues within Congress. The bill language also suggests specific areas of societal and ethical concern that will receive the most attention, at least in the immediate future. My own view of these concerns is first, that they have to be taken seriously, and second, that the scientific community owes the public and Congress a clear and rational vision of nanotechnology that can lead to a productive engagement.

We should begin to construct that clear vision by distinguishing science from science fiction, and by emphasizing nanotechnology's strong links to things we already know a great deal about. While the technologies enabled by atomic scale capabilities are revolutionary, they are not about to spring, like Athena from Zeus, fully armed from the brow of god-like scientists. Nature has experimented with nanostructures since the earth began to cool four-and-a-half-billion years ago, and has blessed us with a rich legacy of examples to stimulate our imaginations. These range from the microstructures of minerals to the intricate molecular mechanisms of life. While it is now possible for us to manufacture structures that do not occur in nature, we are strongly guided by the immense variety of those that do occur. Some of the most important applications of biotechnology are likely to be the tuning up of useful cellular machinery that Nature has not yet had time to evolve to its most efficient form. We have been doing something similar for a century and a half with organic molecules—dyes, for example, or synthetic fibers—and Japanese metallurgists were inventing new microstructures over a much longer history to create edged tools and weapons of legendary quality. They were not aware of the nanoscale origins of their products, but they were producing them just the same.

And throughout this long history, society has built up systematic ways of protecting itself against the undesirable consequences of these evolving technologies. During the past half-century, in particular—and as a direct result of growing scientific knowledge—society has acted through its governmental machinery to establish procedures to protect public and environmental health from new materials technology, whether biological, chemical, or radiological. The 25-year-old RAC process (RAC stands for "Recombinant DNA Advisory Committee"), for example, or a modified version of it recently proposed by an NRC committee chaired by MIT's Stanley Fink, are designed basically to address concerns about new nanoscale phenomena. The Toxic Substances Control Act governing the review and registration of potentially toxic chemicals originated at about the same time as the famous Asilomar Conference that recommended the RAC.

Congress clearly wants to know whether these mechanisms, or reasonable extensions of them, are adequate for responding to concerns about the products of nanotechnology. It is clear that some such products are already covered by existing mechanisms. Can we identify the manner in which new nanotechnology products differ from these older threats? It is important that we do so. I believe the differences are likely to occur in very well-defined areas, and that even in those cases the existing means for addressing threats they may pose to the environment or public health are likely to suffice with relatively little modification or extension.

This emphasis on the continuity of nanoproducts with natural or older man-made substances may help us refocus public attention on the most likely short-term issues.

The media, Hollywood, and some imaginative commentators have focused on self-replicating "nanobots" as the archetypal hydra-headed nano-thing. In my opinion, that is utterly wrongheaded. The most common nanosubstances will be passive structures, suspended or dispersed within a matrix. The most common objective of a nanotechnology project is likely to be the preparation of a bulk functional material or extended structure with nanoscale intrinsic architecture. The production of stand-alone nano-entities will be far down the line, and these will be closely similar to, but simpler than, the intricate naturally occurring proteins, nucleic acids, and the cellular machinery comprising them. For many years, biotechnology will remain far ahead of nanotechnology in producing new entities of this sort, and I think it likely that the protective protocols developed for biotechnology will suffice for hazard control. The ethical issues associated with human biological applications of nano-products are the same as with similar applications of "genetically engineered" bio-products. I am not saying we have answered all ethical questions that are raised by such possibilities as the sensory enhancement and protracted longevity promoted by these applications, but the idea that there are procedures already in place to deal with these new applications ought to be reassuring.

Nanoparticles of chemical substances have properties that differ from the bulk. Probing and understanding those differences are part of the exciting unfinished work of nanoscience. Perhaps our system of cataloguing chemicals needs to be extended to account for the spectrum of new characteristics, analogous to nuclear isotopes that appear at the nanoscale. It seems unlikely to me, however, that the current system for identifying, registering, and controlling hazardous chemicals will need to be changed very much to accommodate this new category of substances.

Let me close by pointing out that ethical decisions about the introduction of new things or new processes can be arranged in two broad categories that ought to be kept separate. Decisions based on the sacredness of objects, entities, or conditions fall in one category. In the other are decisions based on potential harm to individuals or societies, including the environment that sustains all life. The former are easier for science, but much more difficult for society to deal with. The latter category depends heavily on science to assess potential harm. But I believe that neither category is new, nor requires essentially novel kinds of thinking; only a particularizing of principles to the case at hand.

This workshop is important, as was the one that preceded it in 2000. The organizers have brought together experts from diverse fields, and I encourage you to take advantage of this opportunity to share ideas with others who are thinking about the uses and implications of nanotechnology. I am grateful to all of the participants for devoting their time and their ideas to the elucidation of these thorny topics, and look forward to the proceedings.

2. Introductory and Summary Comments

Reference

1. 21ˢᵗ Century Nanotechnology Research and Development Act, 108th Congress, 1st sess., Dec. 3, 2003, Public Law 108–153, 15 USC 7501, Page 117 Stat. 1923, Washington, DC: United States Government Printing Office (2003).

PREPARING THE PATH FOR NANOTECHNOLOGY

Phillip J. Bond, [former] Undersecretary for Technology, U.S. Department of Commerce

We come together at an exciting and dramatic time for all of us engaged in the promotion, research, and development of nanotechnology. It has only been a few years since the United States launched the National Nanotechnology Initiative (NNI)—a bold, visionary effort to harness the extraordinary power of matter at the atomic level. So it's truly amazing how far we've come in so short a time—lifting nanotechnology out of the genre of science fiction, into our academic and industrial laboratories, and more recently, into the marketplace.

And now thanks to President Bush and bipartisan support in Congress, the NNI has been formally enacted through passage of the 21ˢᵗ Century Nanotechnology Research and Development Act. I want to offer my congratulations and thanks to the NanoBusiness Alliance, most of you in this room, and to the others who contributed to bringing this important piece of legislation to fruition.

The Act puts the President's National Nanotechnology Initiative into law and authorizes $3.7 billion over the next four years for the program, and that doesn't even include investments that will be made by the Departments of Defense, Homeland Security and the National Institutes of Health. The legislation establishes a coordination office, advisory committees, and regular program reviews to ensure that taxpayer money is being spent wisely and efficiently. In addition, the bill mandates the establishment of research centers and education and training efforts, and charges the Commerce Department with several things, including the following:

- research at our National Institute of Standards and Technology (NIST) to support nanotechnology metrology, reliability and quality assurance, processes control, and manufacturing best practices

- disseminating research results to small and mid-size manufacturers through our Manufacturing Extension Partnership

- establishing a clearinghouse of information related to the commercialization of nanotechnology research, including

  – providing information on regional, state, and local commercial nanotechnology initiatives

- aiding in the transition of Federally funded R&D into commercial and military products

- identifying and promoting best practices by government, universities, and private sector laboratories in moving technology to market

- identifying ways to overcome barriers and challenges to technology deployment

- conducting analysis on nanotechnology's impact on manufacturing and workforce

And apropos of our workshop this week, the Act mandates the establishment of a center for, and research into, the societal and ethical consequences of nanotechnology.

First Message: Nanotechnology is coming and it won't be stopped

I begin my remarks this afternoon with a single ironclad belief: nanotechnology—with its myriad evolutionary and revolutionary applications—is coming, and it can't be stopped.

Some voices around the world are calling for a slowdown or even an outright moratorium on nanotechnology research and development. To those calling for a slowdown or halt on the nanotechnology front, I say instead: Prepare for the inevitability of a world blessed with nanotechnology and nano-enabled products and services. The economic promise, the societal potential, and the human desire for rolling back the frontiers of knowledge—to go where no one has gone before—are forces that cannot be held back.

In every corner of the globe, research is underway to unlock the secrets and unleash the power of the nanoscale universe. With each new dawn comes powerful new insights. I would speculate that not a day goes by now in which an important new nanotechnology discovery is not made.

Today, large wealthy industrial nations and smaller industrializing nations aspiring to wealth are making significant investments in nanoscale science, technology, and engineering in the anticipation of reaping economic and societal rewards. And these investments are increasing rapidly.

In 1997, global public investment in nanotechnology was less than $500 million. Today, more than 50 nations have nanotechnology programs. This year, I would estimate the public investment to be on the order of $3 billion. And that's just the public investment alone. In the United States, private R&D exceeds public R&D by more than two to one. If this holds true in nanotechnology, then private sector R&D in the United States alone could be approaching $2 billion. With this level of investment, nano-based innovation is inevitable.

Second Message: Even if it could be stopped, it would be unethical to stop it

Those who would have us stop in our tracks argue that it is the only ethical choice. I disagree. In fact, I believe a halt, or even a slowdown, would be the most unethical of choices. Setting aside for a moment all of the economic value that nanotechnology holds—and with it, the ability to improve people's standard of living, healthcare, nutrition, etc.—let's look at a few of the things nanotechnology offers as possibilities for societies and individuals, especially in nano's convergence with other enabling technologies:

- freedom from pollution through clean production technologies

- the ability to repair existing environmental damage

- the ability to feed the world's hungry

- the ability to enable the blind to see and the deaf to hear

- the ability to eradicate diseases and to offer protection against harmful bacteria and viruses

- the ability to extend the length and the quality of life through the repair—and eventually even the replacement—of failing organs

The list goes on. So, I ask: Given the promise of nanotechnology, how can our attempt to harness its power at the earliest opportunity—to alleviate so many of our earthly ills—be anything other than ethical? Conversely, how can a choice not to attempt to harness its power be anything other than unethical?

Third Message: The United States not only leads the world in nanotechnology R&D, but in addressing associated societal and ethical issues

As a representative of the U.S. Department of Commerce, many of the speeches I deliver on nanotechnology focus on the question of the U.S. leadership position in nanotechnology. It may not surprise you that, in my view, the United States is unquestionably the global leader in nanotechnology research, development, and commercialization.

If you look at the numbers, whether in patents or publications, the United States is far ahead. And we seem to have in our national genetic code something that makes us serial entrepreneurs. Many countries can compete with us in per capita patents, but few can compete with us in getting patents to market.

But we are also the world leader in addressing the prospective societal and ethical issues associated with the development and commercialization of nanotechnology.

The NNI has, since its inception, included a focus on these issues in tandem with our R&D agenda. The United States began examining these issues sooner and has made more investments than any other nation.

- Our efforts go back more than a decade. The National Science Foundation (NSF) has supported research on the health effects of nano-sized particles since 1991 as part of its Nanoparticle Synthesis and Processing program.

- This year NSF alone will spend more than $25 million on societal, ethical, and educational issues related to nanotechnology, and an additional $33 million on environmental and health implications.

- In addition, each of the NSF-funded Nanoscale Science and Engineering Centers must apply a portion of its budget to addressing societal, ethical, and educational issues.

- And finally, as demonstrated here today, the White House National Science and Technology Council's interagency subcommittee on Nanoscale Science, Engineering, and Technology (NSET) is also taking on the challenge, convening representatives from government, industry, and academia to proactively address these issues. In addition to today's workshop, last month NSET hosted a workshop on Nanotechnology and the Environment: Applications and Implications.

- And, of course, Federal officials—such as Mike Roco, Clayton Teague, and I—are using the near-continuous stream of nanotechnology business and technology conferences to raise awareness of the need to tackle these issues and to invite participation in this vital dialogue.

Fourth Message: Technology is a two-edged sword

Technology has always been a two-edged sword—offering both upsides and downsides. It has the potential to be used for good or for ill. Sometimes, even when we have attempted to put technology to use for good, it has had unexpected negative consequences.

But the history of human progress is the history of our ability to exploit the benefits of technology while effectively identifying, addressing, and minimizing its downside.

Throughout the 20th century, as technology became more powerful and advanced more quickly, we learned to hone the useful edge of technology's blade, while dulling its other edge—exploiting nuclear power for clean energy, while building a global regime to prevent its proliferation, for example.

Fifth Message: Rapid advance of revolutionary technologies can create ethical and societal challenges beyond our current framework

The technologies under development today—especially the converging technologies of nanotechnology, biotechnology, information technology, and cognitive sciences—are so powerful and revolutionary, their applications are likely to create ethical and societal challenges beyond our current framework. Here are some examples:

- Smart dust sensors, bacteria-size microprocessors, extreme high-density data storage devices, and super-high-speed data communications—embedded in materials, devices, clothing, structures, almost everywhere—could create the ultimate interconnected environment. That kind of utility—instant, untethered access to all the knowledge the world has to offer—would provide enormous personal and societal benefits. For example, as a parent, I might like to know where my daughters are every second. But think about what that would mean for society at large. For an American society, which values privacy, there are some real issues there.

- Or what about issues associated with what *Wired* magazine called the emerging new race it dubbed *robo-sapiens*. What does it mean to be a human being? Do embedded appliances transform me into a post-human creature? I'm 47, and I'm having trouble pulling up names. I'd like to be able to plug in a little additional memory sometimes. And as I was putting in my contacts this morning I thought maybe it's not so bad to trade in the old 20-20 for some infrared or telescopic, microscopic, whatever, and have enhanced eyesight. How far can we go? How far should we go? Does any of this change our essential humanness?

- Okay, so maybe adding electronic appliances tied into our brains may not change your concept of humanness; then what about the customization of our DNA? What if we were to add plant or animal DNA to our own code, for whatever useful benefits it might yield—resistance to disease, night vision, improved strength, and endurance?

- It's certainly conceivable. Now you are no longer talking about improvements, but changes to our fundamental human blueprint. Should it be allowed? Is it a matter of individual choice? Should it be regulated? Who should regulate—the Food and Drug Administration (FDA), the U. S. Department of Agriculture (USDA), the United Nations? What if a nation decides to mandate such enhancements to provide a commercial competitiveness or military advantage? This represents a new area for moral and ethical consideration.

- And what if we allow it? We talk a lot about the division in the world and in specific localities and regions between the haves and have-nots. Think about a world of nanotechnology haves and have-nots: people who may acquire enhanced cognitive abilities, whose physical abilities may be enhanced, and who may be nourished by foods designed to knock out diseases that are killing others around the world. Think of the new moral, ethical, and societal issues this raises.

Sixth Message: Revolutionary technologies can create public apprehension and fear, resulting in efforts to stop their advance

Given these moral and ethical challenges, it is not surprising that such revolutionary technologies create a sense of apprehension and fear:

- Disruptive technologies can transform or eliminate entire industries and occupations, leading to the loss of one's job—and its accompanying hardships—as well as a shifting of economic power and opportunity among nations, regions, and localities.

- Sometimes these technologies fall into the hands of those who do not have good intentions. We have seen the horrible consequences of what even a common technology can do in the hands of terrorists. The thought of what such people could do with more advanced technology is truly frightening. Here are some recent examples of what has already been done:

  - A team of biologists recently created a polio virus in vitro "from scratch."

  - Researchers recently inadvertently published a technique that could be used to enhance the virulence of pathogens, such as anthrax or smallpox, greatly increasing their lethality.

  - Scientists have synthesized a key smallpox viral protein and shown its effectiveness in blocking critical aspects of the human immune response.

- Then there are the missives from respectable authors that raise even more profound concerns. Take, for example, Bill Joy's infamous piece about how the future doesn't really need us.

Faced with such fears about the impact of new technologies, people throughout history have sought to stop their advance:

- During the Industrial Revolution, Dutch workers threw their shoes—sabots—into the machinery in an attempt to damage the technology that they believed would take their jobs.

- Automobiles faced early opposition. When they first became available, some cities banned them. San Francisco had a law that mandated parking your car at the edge of the city and riding a horse or carriage into town.

- Thomas Edison attempted to use such fear to manipulate the public for his own financial benefit. With a vested interest in the success of direct current, Edison sought to undermine the use of alternating current by holding public demonstrations of its danger by electrocuting animals—dogs, cats, horses, even an elephant.

- "If man were meant to fly..." was a common refrain raised by the fearful and skeptical in opposition to commercial aviation.

The same technologies that have brought scientists together—I'm thinking information technology, in particular—have made our world smaller and have brought more people into the public square, people driven by their fears and concerns about the technology under development. And not only are more people involved, they are essentially looking over your shoulder, watching what you are doing in near real-time.

Seventh Message: The body politic is susceptible to the virus of fear

We also know from history that the body politic is susceptible to the virus of fear. When the public catches a public-policy cold virus, their elected representatives sneeze. Our democratic institutions are designed to be responsive to the public. To keep technology moving forward, we must prevent fear from taking hold among the public.

Eighth Message: We must identify legitimate ethical and societal issues and address them as soon as possible

So we can't afford to wait to deal with these things. We need to wrestle with them now.

The first thing we need to do is to sort legitimate concerns from imaginary ones, those that are based on science from those based in science fiction. Then we must debunk and dismiss the latter and devote time, attention, and resources to seriously address the former.

We cannot allow ourselves—or the public—to be distracted or misled by capricious claims, foundationless fears, wanton warnings, pompous pronouncements, and arbitrary assertions. We must devote our efforts to addressing the legitimate concerns.

One reason we can't afford to wait is because the public policy apparatus does not move quickly. It is not designed to move quickly. It is a very different environment than the dynamic, fast-changing one in which you work. So to engage effectively in the political arena, you must think and act far ahead.

Ninth Message: We need a holistic approach, with scientists and engineers playing a key role

To effectively address these questions, the NNI recognizes, we need a holistic approach that embraces ethicists, philosophers, theologians, historians, consumer advocates, business leaders, public officials, and others, with scientists and engineers playing a unique and critical role.

Scientists and engineers are in the best position to contribute to sound policy development, addressing legitimate concerns and allaying irrational public fear. Scientists and engineers alone have the scientific and technical knowledge necessary to sort the wheat from the chaff.

In addition, while not historically great communicators, scientists and engineers have unique credibility with the public in speaking to these issues. We need to communicate frequently, clearly, and proactively with the public about nanotechnology to ensure that Americans have all of the knowledge they need—complete and balanced—to make reasoned judgments on these issues, and scientists and engineers must play a central role in this effort.

Tenth and Final Message: Addressing societal and ethical issues is the right thing and the necessary thing

Finally, I want to leave you with this thought. Addressing societal and ethical issues is *the right thing to do* and *the necessary thing to do*. It is *the right thing to do* because as ethically responsible leaders we must ensure that technology advances human well-being and does not detract from it. It is the *necessary thing to do* because it is essential for speeding technology adoption, broadening the economic and societal benefits, and accelerating and increasing our return on investment.

Under the leadership of Secretary Don Evans, the Commerce Department has adopted the theme "American Jobs, American Values." While exploring and dealing with societal and ethical issues concurrently with our development and commercialization of nanotechnology, we can and must achieve both: creating American jobs, while honoring and upholding American values!

The good news is that throughout history, we have successfully managed the downsides of technology—often through great effort—while enjoying the extraordinary benefits it yields. Nanotechnology should be no exception.

This conference is one more critical step down that path. Thank you for your contribution to this important work.

NNI AFTER THREE YEARS (2001-2003): SETTING NEW TARGETS FOR RESPONSIBLE NANOTECHNOLOGY[†]

M.C. Roco, National Science and Technology Council, National Science Foundation

The National Nanotechnology Initiative (NNI) is a visionary research and development program that coordinates 23 departments and independent agencies; the total investment in fiscal year (FY) 2004 was about $1 billion. The program started formally in FY 2001 (October 2000) and was the result of the bottom-up proposal of an interagency group on nanoscale science and engineering that got started in 1996 [1, 2, 3]. The Federal nanotechnology investment per agency since the beginning of NNI is given in Table 2.1.

The main goals of the NNI are to:

* Maintain a world-class research and development program aimed at realizing the full potential of nanotechnology

* Facilitate transfer of the new technologies into products for economic growth, jobs and other public benefit

* Develop educational resources, a skilled workforce, and the supporting infrastructure and tools to advance nanotechnology

* Support responsible development of nanotechnology

Indeed, nanotechnology's shift in focus from the microscale to the molecular and nanoscale will be essential for future advances in both the digital revolution and modern biology—and may change the very foundation of education, medicine, manufacturing, and the environment. Initially, the NNI was driven by science as outlined in "Nanotechnology Research Directions" [4], but after 2002, technological innovation rose in importance. Industry has become a strong supporter and its long-term R&D nanotechnology investment is expected to surpass the Federal NNI expenditures next year. Also, more than 20 states in the United States have realized that nanotechnology has economic potential and made multi-annual financial commitments in 2003 to nanotechnology that total more than half the NNI annual budget. The worldwide government investment in nanotechnology (in part stimulated by the NNI) is about $4 billion in 2005, a ninefold increase as compared to about $430 million in 1997 (Table 2.2, Fig. 2.1).

Nanotechnology is expanding in a natural and robust way. We are creating the systematic control of matter at the nanoscale. We have clear research and education needs in the national and international context. The White House and Congress have recognized the importance of nanotechnology in the future of the United

---

[†] This presentation and accompanying charts and tables have been updated by the author since the 2003 workshop.

Table 2.1
Contribution of Key Federal Departments and Agencies to NNI Investment*

Federal Department or Agency	FY 2000 Actual ($M)	FY 2001 Actual ($M)	FY 2002 Actual ($M)	FY 2003 Actual ($M)	FY 2004 Actual ($M)	FY 2005 Estimate ($M)	FY 2006 Request ($M)
National Science Foundation (NSF)	97	150	204	221	256	338	344
Department of Defense (DOD)	70	125	224	322	291	257	230
Department of Energy (DOE)	58	88	89	134	202	210	207
National Institutes of Health (NIH)	32	40	59	78	106	142	144
National Institute of Standards and Technology (NIST)	8	33	77	64	77	75	75
National Aeronautics and Space Administration (NASA)	5	22	35	36	47	45	32
National Institute for Occupational Safety and Health (NIOSH)	-	-	-	-	-	3	3
Environmental Protection Agency (EPA)	-	5	6	5	5	5	5
Homeland Security (TSA)	-	-	2	1	1	1	1
Department of Agriculture (USDA)	-	-	-	1	2	3	11
Department of Justice (DOJ)	-	1	1	1	2	2	2
TOTAL	**270**	**464**	**697**	**863**	**989**	**1,081**	**1,054**
(% of FY 2000)	(100%)	(172%)	(258%)	(320%)	(366%)	(400%)	(390%)

* Each Fiscal Year (FY) begins October 1 of the previous calendar year and ends September 30 of the cited year.

States through the *NNI Supplement to the President's FY 2004 Budget* [5] and the *21st Century Nanotechnology Research and Development Act* [6]. The NNI, in collaboration with other worldwide nanotechnology programs, has the potential to bring broad societal changes, from increasing productivity in manufacturing to extending the quality of life and sustainability limits on Earth.

Results of the NNI Investment

There are major outcomes after the first three years (fiscal years 2001–2003) of the NNI. The NNI has already created a nanoscale science and engineering

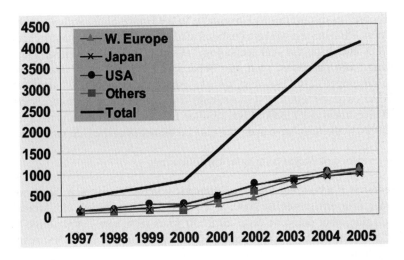

Figure 2.1. International Context—Government nanotechnology R&D investments in the past eight years, 1997-2005.

Table 2.2

Estimated Government Nanotechnology R&D Expenditures, 1997-2005 ($ Millions/Year)

Region	1997	1998	1999	2000	2001	2002	2003	2004	2005
W. Europe	126	151	179	200	~225	~400	~650	~950	1,050
Japan	120	135	157	245	~465	~720	~800	~900	950
USA*	116**	190**	255**	270**	465**	697**	863**	~989	1,081
Others	70	83	96	110	~380	~550	~800	~900	1,000
Total	432	559	687	825	1,535	2,367	3,113	3,739	4,081
(% of 1997)	(100%)	(129%)	(159%)	(191%)	(355%)	(547%)	(720%)	(866%)	(945%)

Explanatory notes: Estimates include research in nanotechnology as defined by the NNI (this definition does not include MEMS, microelectronics, or general research on materials), and reflect the publicly reported government spending.

"W. Europe" includes countries in EU (15) and Switzerland. Rates of exchange $1 = 1.1 Euro until 2002, = 0.9 Euro in 2003, and = 0.8 Euro in 2004-2005. National and EU funding are included.

Japanese rate of exchange $1 = 120 yen until 2002, = 110 yen in 2003, = 105 yen in 2004-2005.

"Others" includes Australia, Canada, China, Eastern Europe, former Soviet Union, Israel, Korea, Singapore, Taiwan, and other countries with nanotechnology R&D.

* A fiscal year begins in USA on October 1, six months before most other countries.

** Denotes the actual expenditures recorded at the end of the respective fiscal year.

"powerhouse" of discoveries and inventions in the United States with about 40,000 researchers, students and workers qualified at least in one aspect of nanotechnology. The R&D landscape for nanotechnology research and education has changed, advancing from questions, such as, "What is nanotechnology?" and "Could it ever be developed?" to "How can we take advantage of it faster?" and "Who is the leader?" The FY 2005 NNI investment is about four times the corresponding Federal investment in FY 2000 ($1081 million from $270 million), and the attention is extending to the legislative and even judicial branches of the U.S. Government.

Further evidence of progress includes the following:

- Research is advancing towards systematic control of matter at the nanoscale faster than envisioned in 2000, when NNI was introduced with words like "Imagine what could be done 20 to 30 years from now." After three years, in 2003, the NNI supports about 2,500 active awards in about 300 academic organizations and about 200 small businesses and nonprofit organizations in all 50 states. The time of reaching commercial prototypes has been reduced by at least a factor of two for several key areas such as detection of cancer, molecular electronics, and special nanocomposites. About half of highly cited papers by the Institute for Scientific Information, High-Impact Papers electronic database appear to continue to originate from the United States [7].

- Systemic changes are being implemented for education, by earlier introduction of nanoscience and reversing the "pyramid of science" with the understanding of the unity of nature at the scale of atoms, molecules, and their assemblies taught in the earliest years of science education. In 2002, NSF announced the nanotechnology undergraduate education program, and in 2003, the nanotechnology high school education program. In the next years, we plan to change the language of science even earlier and involve science museums to "seed" that language in K-12 students. About 7,000 students and teachers have been trained in 2003 with NSF support. All major science and engineering colleges in the United States have introduced courses related to nanoscale science and engineering in the last three years.

- Significant infrastructure has been established in more than 60 universities with nanotechnology user capabilities. Five networks (Network for Computational Nanotechnology, National Nanotechnology Infrastructure Network, Oklahoma Network for Nanotechnology, the DOE large facilities network, and the NASA nanotechnology academic centers) have been established.

- Industry investment has reached about the same level of investment as the NNI in long-term R&D, and almost all major companies in traditional and emerging fields have nanotechnology groups at least to survey the competition. For example, Intel has reported $20 billion revenues from nanotechnology in 2003. About 61 percent of patents (about 1,011 of 1,644) related to nanotechnology (searched by keywords in the title, abstract, and claims) as recorded by the

U.S. Patent and Trade Office in 2002 are from the United States, while the NNI funding represents about 25 percent of the world government investment (about $770 million of $3.0 billion). According to a NanoBusiness Alliance estimate of business activity in late 2003, 70 percent of the nanotechnology start-up companies worldwide (approximately 1,100 of 1,500) were located in the United States. Despite the general economic downturn, nanotechnology venture funding in the United States doubled in 2002 as compared to 2001. NSF supported more than 100 small businesses with an investment of $36 million between 2001 and 2003.

- The NNI's vision of a "grand coalition" of academe, government, industry, and professional groups is taking shape. More than 22 regional alliances have been established throughout the United States to develop local partnerships and support commercialization and education. Professional societies have established specialized divisions, organized workshops and continuing education programs. Among the professional societies are the American Association for the Advancement of Science, American Chemical Society, American Physics Society, Materials Research Society, American Society of Mechanical Engineers, American Institute of Chemical Engineers, Institute of Electrical and Electronics Engineers, and American Vacuum Society.

- Societal implications were addressed from the start of the NNI, beginning with the first research and education program on environmental and societal implications, announced in a program solicitation by NSF in July 2000. In September 2000, the report on *Societal Implications of Nanoscience and Nanotechnology* was issued. In 2004, the number of projects in the area grew significantly, funded by NSF, EPA, NIH, DOE, and other agencies. Awareness of potential unexpected consequences of nanotechnology has increased, and Federal agencies meet periodically to discuss those issues.

Where Is the NNI Going from Here?

Nanotechnology has the potential to change our comprehension of nature and life, develop unprecedented manufacturing tools and medical procedures, and even change societal and international relations. The first set of nanotechnology grand challenges was established in 1999–2000. Let's imagine again what could be done. I envision several potential developments by 2015:

- Half of the newly designed advanced materials and manufacturing processes are built using control at the nanoscale. Even if this control may be rudimentary as compared to the long-term potential of nanotechnology, this will mark a milestone towards the new industrial revolution outlined in 2000. By extending the experience with information technology in the 1990s, I would estimate an overall increase in productivity of at least 1 percent per year because of these changes. Ahead are several challenges. Visualization and numerical simulation of three-dimensional domains with nanometer resolution will be necessary for engineering applications. Nanoscale-designed catalysts will expand the use in "exact" chemical manufacturing to cut and link molecular assemblies with minimal waste. Silicon transistors will reach dimensions smaller than 10 nm and will be integrated with molecular or other kind of nanoscale systems (beyond or integrated with Complementary Metal-Oxide Semiconductor technology [CMOS]). Changing our goals and strategies in this area is the experimental proof of concept, completed in 2003, which showed that CMOS can work at 5 nanometer gate lengths (and potentially at a smaller scale). One may recall that in 2000, we contemplated the "brick wall" of physical principles that would limit the advancement of silicon technology by the end of this decade. Now we are looking to advances in CMOS technology over another decade (by 2020) and then to its integration with bottom-up molecular assembling.

- Suffering from chronic illnesses is being sharply reduced. It's conceivable that by 2015, our ability to detect and treat tumors in their first year of occurrence might sharply reduce suffering and death from cancer. In 2000, we aimed for earlier detection of cancer within 20 to 30 years. Today, based on the results obtained during the period 2001 through 2003 with respect to understanding the processes within a cell and new instrumentation to characterize those cellular processes, we are trying to eliminate cancer as a cause of death if treated in a timely manner. Pharmaceutical synthesis, processing, and delivery will be enhanced by nanoscale control, and about half of pharmaceuticals will use nanotechnology as a key component. Modeling the brain based on neuron-to-neuron interactions will be possible by using advances in nanoscale measurement and simulation.

- Converging technologies from the nanoscale will establish a mainstream pattern for applying and integrating nanotechnology with biology, electronics, medicine, learning, and other fields. It includes hybrid manufacturing, neuromorphic engineering, artificial organs, expansion of the life span, and enhanced learning and sensorial capacities. New concepts in distributed manufacturing and multi-competency clustering will be developed.

- Life cycle sustainability and biocompatibility will be pursued in the development of new products. Knowledge development in nanotechnology will lead to reliable safety rules for limiting unexpected environmental and health consequences of nanomaterials. Synergism among life cycles of various groups of products will be introduced for overall sustainable development. Control of nanoparticles will be performed in air, soils, and waters using a national network for monitoring and remediation.

- Knowledge development and education will begin with instruction about the nanoscale instead of the microscale. Earlier exposure to nanoscience education could change the role of science and enhance motivation for schoolchildren. A new education paradigm not based on disciplines but on the unity of nature, and education-research integration will be tested for K-16 (reversing

Table 2.3
Examples of NNI Projects Supporting
Social Implications Research

Project	Agency	Institution
Nanotechnology and its publics	NSF	Pennsylvania State University
Public information and deliberation in nanoscience and nanotechnology policy (SGER)	Interagency	North Carolina State University
Social and ethical research and education in agrifood nanotechnology (NIRT)	NSF	Michigan State University
From laboratory to society: developing an informed approach to nanoscale science and engineering (NIRT)	NSF	University of South Carolina
Database and innovation timeline for nanotechnology	NSF	UCLA
Social and ethical dimensions of nanotechnology	NSF	University of Virginia
Undergraduate exploration of nanoscience, applications and societal implications (NUE)	NSF	Michigan Technological University
Ethics and belief inside the development of nanotechnology (CAREER)	NSF	University of Virginia
All centers, NNIN and NCN have societal implications components	NSF, DOE, DOD, and NIH	All nanotechnology centers and networks

the pyramid of learning [8]). Science and education paradigm changes will be at least as fundamental as those that occurred during the "microscale S&E transition" that originated in the 1950s, when microscale analysis and scientific analysis were stimulated by the space race and digital revolution. Stimulated by nanotechnology products, the new "nanoscale S&E transition" will change the foundation of analysis and the language of education. This new "transition" originated at the threshold of the third millennium.

- Nanotechnology businesses and organizations will restructure towards integration with other technologies in technology platforms, distributed production, and clusters of complementary activities. Traditional and emerging technologies will be equally affected.

Responsible Development of Nanotechnology

A main reason for developing nanotechnology is to extend the limits of sustainable development. One way to accomplish this is through "exact" manufacturing at the nanoscale with small consumption of energy, water, and materials, as well as minimized waste. Another way to promote sustainable development is through mitigating the effects of existing nanoscale contaminants from traditional activities, such as operating combustion engines or from natural sources, such as biomineralization and sediment fragmentation. Third is controlling the evolution of existing and newly released nanomaterials in the environment. The NNI annual investment in nanoscale research with relevance to the environment was estimated at about $50 million in 2002, of which NSF awarded about $30 million and EPA awarded about $6 million. If one would add the research for societal and educational implications, the investment is about 10 percent of the total annual NNI budget. These efforts are funded by several agencies, including NSF, EPA, NIH, DOE, NIOSH (National Institute for Occupational Safety and Health), USDA, and DOD.

NSF has focused on nanoscale processes in the environment and on societal implications in its programs since August 2000. NSF will award about $16 million in 2004 for grants with a primary focus on environment and nanotechnology, and additionally about the same amount for multidisciplinary projects including environmental issues. A list of 100 grants, including abstracts, is available on http://www.nsf.gov/crssprgm/nano/activities/nni01_03_env.jsp. Support for social, ethical and economic implications is growing. Information on two grants of more than $1 million each with a focus on the interaction with the public and the creation of databases is available on http://www.nsf.gov/od/lpa/news/03/pr0389.htm. NSF's Nanoscale Science and Engineering Centers (NSEC) and the National Nanotechnology Infrastructure Network (NNIN) are required to have research and education components addressing environmental and societal implications.

Three Federal agencies now have focused efforts to study the potential risks of exposure to nanomaterials: the National Toxicology Program (NTP)—a multiagency effort established in the Department of Health and Human Services—NIOSH, and EPA. The NTP studies will focus on the potential toxicity of nanomaterials, beginning with titanium dioxide, several types of quantum dots, and fullerenes. The first studies will examine distribution and uptake by the skin of titanium dioxide, fullerenes, and quantum dots. The NTP is also considering conducting inhalation studies of fullerenes and is exploring ways to assist NIOSH in the development of protocols for research on inhalation potential and effects of carbon nanotubes. The NIOSH provides research, information, education and training in the field of occupational safety and health. In 2004, NIOSH initiated several research projects focused on nanotechnology, including a five-year program to assess the toxicity of ultrafine and nanoparticles.

EPA is funding research at universities to examine the toxicity of manufactured nanomaterials such as quantum dots, carbon nanotubes, and titanium dioxide. In addition, current and past work in ultrafine particulates at EPA labs and funded through the extramural program at EPA provides information on the effects of nanoparticles on human health. Scientists funded by the NIH also are studying the chemistry, biology, and physics of nanoscale material interactions at the molecular and cellular level through *in vitro* and *in vivo* experiments and simulation models.

The Department of Defense is supporting a Multidisciplinary University Research Initiative (MURI) program to create predictive models for cellular response to nanoparticles of varying size, shape, charge, and composition and their influence on the cellular, sub-cellular, and biomolecular levels. This research is creating a significant body of knowledge of reactions between nanoscale materials and biological materials.

All material stuff around us, either natural or man-made, has structure at the nanoscale. All living cells, for instance, interact with nanostructures when they feed, breed or are touched by viruses. Thus, facilitated by new investigative methods, development of knowledge at the nanoscale is a natural trend in science and engineering. This knowledge may prepare us to address unexpected risks of human activity, such as encountering unknown viruses and bacteria. The knowledge also might help us to address challenges raised by nanostructures themselves, particularly new functions of the same chemical composition and more reactive surfaces of nanostructures.

NNI research is developing new knowledge regarding environmental, health and safety issues through more than 120 projects underway at the end of 2003, including several centers at the University of California, Davis (nanoparticles in the environment); Worcester Polytechnic Institute (air pollution); University of Illinois at Urbana-Champaign (water purification); Rice University (nanostructures in the environment); and University of Notre Dame (nanoparticles in soils). Researchers are addressing such questions as: What is different about artificially created nanostructures? How would those nanostructures behave differently from bulk if released in the environment? Nanotechnology will develop in the areas where potential advantages will exceed the impact of potential risks and where the potential risks are limited and can be addressed. Current approaches are attempting to address nanotechnology impacts in research or production within the existing system applications such as biology, chemistry or electronics.

The key questions asked by technology users and the public concern economic development and related issues, such as commercialization, education, infrastructure, and environmental, health, ethical, and legal aspects. We have the responsibility to increase productivity, better use natural resources, reduce poverty and hunger, improve healthcare, and enhance human resources. We also must address health and environmental risks and related efforts to reduce them. The public policy response must be balanced between public benefit and risk. Consideration of the opinions of individual groups—at times different from the largest majority and sometimes conflicting with scientific facts—must occur in the context of broader societal goals.

The vision of few intelligent nanometer robots mentioned in science fiction literature, for example, the novel *Prey* by Michael Crichton, leads to immediate criticism by some groups that are concerned that such robots would take over the world and damage the environment. This criticism ignores input from researchers who note that basic laws of mass and energy conservation may not lead to infinitely multiplying material objects, and that only a complex system of presumably already known living systems may be able to multiply and be intelligent.

The government's role is to provide R&D support for knowledge development, identify possible risks for health, environment, and human dignity, and inform the public with a balanced approach about the benefits and potential unexpected consequences of nanotechnology.

The NSF prepared a report entitled *Societal Implications of Nanoscience and Nanotechnology* in September 2000 and published it for broader public distribution in 2001 [9]. The proceedings were followed by various program solicitations and the assignment to the National Nanotechnology Coordination Office (NNCO) in 2001 of a monitoring role for potential unexpected societal implications. The NNCO also has the role of communicating with the public.

In 2003, a subgroup of the NSET Subcommittee, the Nanotechnology Environmental and Health Implications (NEHI) working group, was established to address environment, health, and safety (EHS) issues. Among those issues are identification and prioritization of EHS research needs and communication of information pertaining to the EHS aspects of nanomaterials to researchers and others who handle and use nanomaterials.

In another follow-up to the 2000 *Societal Implications* report, NSF has made support for social, ethical, and economic research studies a priority by (a) including this as a new theme in the NSF annual program solicitations since 2000; (b) requiring its nanotechnology research and education centers to address societal implications of the research performed in the respective center; and (c) conducting a study on the impact of technology and converging technologies from the nanoscale [10].

NSF has pursued the research and education themes "Nanoscale processes in the environment" and "Societal and Educational Implications of Nanotechnology" as part of its NNI programs since July 2000 (annual program solicitations NSF 00-119, 01-157, 02-148, 03-043, 03-044), and 100 examples of awards made in this area are posted on www.nsf.gov/nano, listed under Solicitations and Outcomes. Examples of projects supporting societal implications are given in Table 2.3. EPA has had annual program announcements in the STAR program with focus on nanotechnology and the environment since 2002; in FY 2003, 22 awards were made and, in 2004, about 12. DOE has included nanoscience in environmental research performed at several National Laboratories, such as Oak Ridge in Tennessee and the Environmental Molecular Sciences Laboratory in Washington State. Additional Small Business Innovation and Research/Small Business Technology Transfer Program (SBIR/STTR) awards were made at NSF after 1999 when nanotechnology was specifically targeted in the respective program announcements. EPA will have an SBIR solicitation on "Nanomaterials and Clean Technology" with a deadline in May 2004. FDA, EPA and other regulatory agencies are following very closely the research results.

The NNI annual investment in research and education with relevance to environment has increased progressively since 2000. Other programs dedicated to environmental implications of nanotechnology abroad were announced in March 2003 by the European Community and in November 2003 by Taiwan—about three years after the NSF first called for proposals in that area.

One should not sidetrack the efforts for sustainable development by delaying or halting the creation of new knowledge in the field. At the international "Nanotech 2003 and Future" conference in Japan on February 26, 2003, I made an international appeal to researchers and funding organizations "to take timely and responsible advantage of the new technology for economic and sustainable development, to initiate societal implications studies from the beginning of the nanotechnology programs, and to communicate effectively the goals and potential risks with research users and the public" [11]. Since then, I've had discussions with representatives from the European Commission, Asia-Pacific Economic Cooperation, Switzerland, UK, Taiwan, China, Australia, and other countries about this topic. International collaboration is necessary in a field that does not have borders, where the products are sold internationally, and the health and environmental aspects are of general interest.

Nanotechnology is still in the precompetitive phase in most areas where applications are foreseen, and international collaboration is beneficial. Nanotechnology has the long-term potential to bring revolutionary changes in society and harmonize international efforts towards a higher purpose than just advancing a single field of science and technology or a single geographical region. A global strategy guided by broad societal goals of mutual interest is envisioned.

Appendix: Laws and Regulations that Apply to Nanotechnology Development

On December 3, 2003, the President signed into law the 21st Century Nanotechnology Research and Development Act [6]. A section of that law is dedicated to societal implications.

Congress issues authorization laws and funding appropriations for nanotechnology R&D to Federal agencies participating in NNI each year. The number of participating agencies has increased from six agencies in FY 2001 to 10 agencies in FY 2002 and 22 agencies in FY 2005.

These organizations have primary responsibility for implementing regulations and guidance in areas relevant to nanotechnology materials and products:

- Environmental Protection Agency (EPA)

- Food and Drug Administration (FDA)

- National Institute for Occupational Safety and Health (NIOSH)

- Occupational Safety and Health Administration (OSHA)

- U.S. Department of Agriculture (USDA)

- Consumer Product Safety Commission (CPSC)

- U.S. Patent and Trademark Office (USPTO)

Research to establish the knowledge base that is used by regulatory agencies to inform their decision-making process may be performed by Federal agencies, such as NSF, NIH, NIST, EPA, FDA, NIOSH, OSHA, USDA, DOE, and DOD, or may be performed by industry or other private sector research institutions.

The materials and products based on nanotechnology are regulated today within the existing network of statutes, regulations, rules, guidelines, and other voluntary activities. Nanostructures are evaluated by various groups and in different countries as "chemicals with new uses" or as "new chemicals." In some cases, pre-market review and approval are required (e.g., drugs, food packaging, and new chemical compounds). In other cases, post-market surveillance and monitoring apply (e.g., cosmetics and most consumer products). The existing regulatory network will be modified, if necessary. Examples of regulatory laws and standards applicable to nanoparticles and other nanostructures include the following:

- In the environment (in air, water, soils):

  - Toxic Substances Control Act (TSCA), administered by the EPA

  - Clean Air Act for ultrafine particles, administered by the EPA

  - Waste disposal acts, administered by the EPA

- In the work place (aerosol-based standards based on existing health risk data):

  - Permissible Exposure Limits (PELs), established by the Occupational Safety and Health Administration (OSHA)

- Recommended Exposure Limits (RELs), established by the National Institute of Occupational Safety and Health (NIOSH)

- Threshold Limit Values (TLVs), established by the American Conference of Government Industrial Hygienists (ACGIH)

- Personal Protective Equipment to reduce exposure, established by the OSHA and ASTM (American Society for Testing and Materials)

- Nanoparticles for drugs to be metabolized in the human body, to be used as diagnostics or therapeutic medical devices (such as quantum dots); regulated by the FDA.

- Nanostructured particles/substances to be incorporated into food; the FDA and USDA share regulatory authority (such as food additives, food coloring).

- Substances incorporated into consumer products; regulated by the Consumer Product Safety Commission (CPSC) under the Federal Hazardous Substances Act. A focus is on protection of children, who are more susceptible and who sometimes put objects in their mouth that were not intended for that purpose.

Under NEHI coordination, the EPA, FDA, CPSC, OSHA, NIOSH, NIST, USDA, and other agencies are reviewing existing rules and procedures to determine how to use the existing statutes and regulations to review products of nanotechnology, as these products are developed. Where new nanotechnology products differ from existing products and present unique concerns for the environment or public health, modification or extension of rules will be considered.

References

1. M. C. Roco, R. S. Williams, P. Alivisatos, eds., *Nanotechnology Research Directions*, Washington, DC: NSTC (1999), also Dordrecht: Springer (formerly Kluwer) (2000) (available at http://www.wtec.org/loyola/nano/IWGN.Research.Directions/).

2. NSTC, *National Nanotechnology Initiative: Supplement to the President's FY 2004 Budget*, Washington, DC: NSTC (2003).

3. *21st Century Nanotechnology Research and Development Act*, 108th Congress, 1st sess., Dec. 3, 2003, Public Law 108-153, 15 USC 7501, Page 117 Stat. 1923, Washington, DC: United States Government Printing Office (2003).

4. L. G. Zucker, M. R. Darby, personal communication (2003).

5. M. C. Roco, Converging science and technology at the nanoscale: Opportunities for education and training, *Nature Biotechnology* **21**, 1247–1249 (2003).

6. M. C. Roco, W. S. Bainbridge, eds., *Societal Implications of Nanoscience and Nanotechnology*: National Science Foundation Report, Arlington, VA: National Science Foundation (2000), also Dordrecht: Springer (formerly Kluwer) (2001) (available at http://www.wtec.org/loyola/nano/NSET.Societal.Implications).

7. M.C.Roco, W. S. Bainbridge, eds., Converging Technologies for Improving Human Performance: Nanotechnology, Biotechnology, Information Technology and Cognitive Science: NSF-DOC Report, Arlington, VA: National Science Foundation (June 2002), also Dordrecht: Springer (formerly Kluwer) (2003).

8. M.C.Roco, Broader societal issues of nanotechnology, *J. Nanoparticle Research* **5**, 181–189 (2003).

9. 21st Century Nanotechnology Research and Development Act, Public Law 108-153, 108th Congress, 1st sess., Dec. 3, 2003, 117 Stat. 1923, Washington, DC: United States Government Printing Office, (2003).

10. R. W. Siegel, E. Hu, M. C. Roco, eds., *Nanostructure Science and Technology*, Washington, D.C.: NSTC (1999), also Dordrecht: Springer (formerly Kluwer) (1999) (available at http://www.wtec.org/loyola/nano/).

General References

NSTC, *National Nanotechnology Initiative: The Initiative and Its Implementation Plan*, Washington, D.C.: NSTC (2000) (available at http://www.nano.gov/nni2.pdf).

NSTC, *Nanotechnology—Shaping the World Atom by Atom*, Washington, D.C.: NSTC (1999) (available at http://www.wtec.org/loyola/nano/IWGN.Public.Brochure/).

Additional information may be found on www.nano.gov and www.nsf.gov/nano.

SCIENCE AND EDUCATION FOR NANOSCIENCE AND NANOTECHNOLOGY

George M. Whitesides, Department of Chemistry, Harvard University

In a given period of history, one technology or another may be primarily responsible for the major changes in the world. The great advances before World War II were largely in physics. It may well be that the great advances in the next century come in significant part from biology. But in the period between 1950 and now, change has largely been caused by information technology, and information technology has been enabled by making things small. "Smalltech" is remaking the world.

2. Introductory and Summary Comments

The characteristic of nanotechnology we should keep in mind is that it is a part of what I would call *small technology*, that is, shrinking things that used to fill a room and putting them on your desks and in your pocket. The basic idea of "small" makes the big difference, as first illustrated by microelectronics. The nanoscale is smaller than the microscale by a factor 10^3, a factor that is also worth keeping in mind. If you ask when something becomes fundamentally recognizable as new, it is usually when it differs by a factor of 10^3. Nanotechnology has the potential to bring that factor of 10^3 to the general area of small.

One of the most notable characteristics of nanotechnology is the fact that, because it is new, there has been a great deal of work devoted to making a case for it, and that work is often called *hyperbole*. What society knows, of course, is what it hears, and the relationships between reality and perception are always complicated. Exaggerated perceptions of benefit and risk go hand-in-hand: hyperbole and anxiety. It is time, now, to be rational and bring these two into line. We must line up what is real, and what is more complicated than one might think.

New, top-end technology is vitally important to the United States. The nation has a generic strategy for dealing with societal problems: namely to spend money on them. This can be an effective strategy so long as we have high-end, profitable, high-margin, proprietary technologies to work with. We may now ask what will be the leading technology after biology and information technology. There are many contenders: a fully developed World Wide Web, distributed communications with complete portability, technology for globalization, smalltech, biotech, knowledge technology, applied social sciences, intelligent machines and others. A very important one of these is nanotechnology, and we need to pay close attention to it in order to foster the strengths of the United States.

Social implications have a number of dimensions, each with a different axis. There are implications for creation of wealth and jobs (which are not the same), for the culture of national security, for education and so on. We need to think about each of these issues separately, although they are intimately related. Some implications of a new technology—"smalltech"—may turn out to be extraordinary rather than ordinary, revolutionary rather than evolutionary.

Nanotechnology investment levels are comparable in the United States, Europe and Asia, so the competition among regions and countries is a real horse race. It is not clear who will win. The United States has historically relied very heavily on the coupling of a first-rate university system with large-scale industrial development through a venture capital mechanism. For a variety of reasons, such as understandable conservatism in the wake of the dot-com bubble, the venture capital mechanism is not working in the United States as it has in the past. The venture world right now prefers to take a familiar area of technology, which is fully ready for commercialization, and help it go to market. It is much less interested in developing technology and exploring early-stage applications using private funds. The situation with respect to nanotechnology is particularly complicated and inefficient because this is an infrastructure technology. The product of nanotechnology is not itself a final product, but it goes into something—for example, a computer—that becomes a product. It is hard to understand how to handle it in a venture role, because the development cycle is so long. Another issue is trying to understand what the process is that moves an interesting scientific invention into a large company, when large companies have become primarily manufacturing, marketing and distribution organizations. How you get across the chasm between "research" and "product sales" is a particular problem for nanotechnology.

Past technological revolutions have involved relatively large-scale technologies, such as internal combustion engines, aircraft, electronic circuit boards and clusters of cells in the biology-related technologies. Now we have the ability to pick up an iron atom and put it where we want it. The science is genuinely revolutionary. The questions are going to be: How can one make technologies out of that kind of revolution? Will this revolution be comparable to earlier ones in its overall impact on society?

An important issue to remember, however, is that nanotechnology is developing independently of the fundamentally new science. Probably the single most important immediate area for nanotechnology is going to be in electronics, and the engineers at Intel and NEC and Phillips are doing an absolutely splendid job of bringing their technologies down in size into the <100 nanometer range. The consequences of that downsizing will be very important, and downsizing will come by a normal industrial process, regardless of fundamental nanoscience research.

Apocalyptic views of nanotechnology, such as the fear of "gray goo," are irrational and unfounded. There are, however, applications of nanotechnology that could have a large impact in the long term in areas like information, genomics, and privacy, and thus provide risks as well as benefits to society. For those not heavily involved in sciences, we might want to develop a color-coded classification: (1) green for nanotechnology developments that are benign or good, (2) blue for those that are neutral or mixed, and (3) red for things that need to be thought about.

One outcome of "nano" that is certainly going to happen is very dense, very inexpensive information storage; this outcome will be "blue" or "green" or "red." Memory will become free. We are going to be able to collect and store any amount of information that we can imagine, for almost nothing. This ability has some very good features associated with it, and it has some that really need to be worried about in terms of privacy and use of information. Memory storage is enabled by nanotechnology, but it is not specifically nanotechnology.

In the neutral "blue" category, I would put the fact that nanotechnology is going to create many high-end jobs. It will invigorate capitalism, and capitalism has its strengths and its weaknesses with respect to equity and distribution of benefits. It will strengthen national security, but at a potential cost to privacy. It will certainly improve our understanding of nature, and that can be used both for good and for mischief.

We have seen one technology—electronics—that has been consistently revolutionized by the process of making things smaller. But if you look around the world, you will find everywhere large things that can be made small. Analytical devices for clinical diagnostics, sensors for a wide variety of purposes, new kinds of communication systems, machines—all have the potential for this downsizing by a factor of 10^3 or more; this factor makes them fundamentally different.

An important area that has not been much emphasized in the United States is a potential for what I call *ultra precision manufacturing*. If you can make parts in which you control their shape with very, very high precision, you can change your manufacturing systems in ways to make them more reliable, less expensive and able to do things that are not otherwise possible. The potential to control manufacturing processes with nanoscale precision is fundamentally new.

The ethical problems in smalltech—the "red nano"—that should have highest priority, in my opinion, have to do with information, privacy, and alienation from technology. It is hard for people to be relaxed with technologies, such as nanotechnology, that they cannot see and have a hard time understanding. As a new technology, nanotechnology will also inevitably contribute to the potentially destabilizing separation between the wealthy nations and those who have less.

Revolution or Evolution?

New tools may enable routine manipulation at the nanoscale, arranging matter atom by atom. This is fantastic science—a capability unknown at the moment, with unknown implications. Materials will probably be the first major new type of technology that comes out of nanotechnology. Carbon nanotubes are on the order of 20 or 30 nanometers in diameter. This is a kind of matter that we have not been able to manufacture before and which has some remarkably interesting and potentially useful properties. In a sense, of course, we have been using other kinds of nanometer scale matter for centuries, in the whole area called chemistry, without great excitement. So, then, what is it about the nanotechnology area that has the potential to lead to something really quite interesting? We can manipulate single atoms and small aggregates of atoms for the first time. This is a fundamentally new capability [1, 2, 4, 6, 7, 8, 9].

Work at the range of 1 to 30 nanometers has the potential to lead to new, potentially applied quantum phenomena, involving ideas that are truly non-intuitive and have the potential to do things that are truly revolutionary. I don't know whether these will ever happen, but they *could* be the most important products of nanotechnology.

It is absolutely certain that new materials will be created industrially, including some with ultrahigh surface area, low defects, new properties such as ballistic electron transport in silicon nanotubes, involving hierarchical structures, both bottom-up and top-down.

An area that is going to be very important is understanding the cell, the fundamental unit of biology [3]. The cell is an object that is maybe 10 microns across, filled with things—organelles and internal structures—that are nanoscale in size. So, if you want truly to understand life, you have to understand the cell's internal, nanometer-scale structures. The cellular machinery includes ATPases, chloroplasts, ribosomes and other structures; to explore these structures in the context of the cell, one needs to build sensors smaller than a cell.

Because nanoscale components are so small, you can put very large numbers of them in small volumes; small size leads both to portability and to certain phenomena having to do with complexity that we have not been able to study in the past, and have certainly not also been able to *use*.

Microelectronics will continue getting smaller so long as it makes things cheaper by doing so. The cost of building a microelectronic chip factory—a fab line—keeps increasing. We could imagine one costing $50 billion at some time in the not too distant future, and you have to sell a lot of microprocessors at $100 a piece to get a good return on an investment of $50 billion. The figure of merit for electronics is not really smallness per se, but cheapness. Transistors on a chip are getting smaller and cheaper in an evolutionary way. But when this trend runs out, the question will be whether a new science will go down all the way to the molecular scale and give that extra factor of 1,000? We don't know, but it is a very important issue to think about. We must also think of the other factors that are so important: heat dissipation, power distribution, input/output.

The first transistor was made in 1947, and the first integrated circuit in 1958, so the course of their development began about 50 years ago. A sense of how rapidly technology can advance is seen in the fact that the United States makes on the order of 3 billion transistors per second, today. Nanoscale circuits have just become possible now, and if the same trend can extrapolate 50 years into the future, there really is a potential to do things that are quite astonishing. People have begun to ask: When is it going to happen?

An important general rule is that one invention does not by itself make a new technology. A technology requires a whole shelf full of innovations. So microelectronics was not merely the transistor, but single crystal silicon, photolithography, the integrated circuit, displays, microprocessors and memory, software, the laser, optical fiber, and the World Wide Web. The whole system had to exist before there was a real technology. Biotechnology brings together DNA, sex, the double helix, restriction endonucleases, cloning, expression vectors, protein engineering, polymerase chain reaction, the cell cycle, oncogenes, apoptosis, and much more. Nanotechnology is just at the beginning of the period of fruitful development. There are just a few things on the shelf now—buckytubes, top-down nanofab, colloid chemistry, structural biology, self-assembly, molecular electronics, mechanical genome surgery, sophisticated biomimicry, and synthetic complexity (see Figure 2.2). More will happen rapidly. We must not be impatient because this is a 10 to 20 year process.

There is much to learn about how to think about the nanoworld by looking at biology. Consider the nanoscale motor that turns the "propeller" of a bacterium, which could be the model for nanometer scale machines, made using principles that are totally different from anything that we use now. There are also more immediate

Figure 2.2. Pentium die photograph (courtesy of Molecular Expressions).

bio-nano things that can be done, such as using magnetic resonance contrast enhancement agents in medical diagnosis. These agents contain nanoscale particles that are superparamagnetic. Injected into patients, they can improve the resolution in magnetic resonance imaging.

What are the risks specific to nanotechnology? There are many benefits, but we always tend to focus on the risks for good reason. One legitimate risk is the fact that we do not understand in detail the health characteristics of very small particles. Most of them are not going to be a problem, but those that are not metabolized may be an issue. The same kinds of particles, in the same range of sizes, are of concern in the so-called black carbon problem, including the soot that comes out of the exhaust of a diesel engine, smoke from fires and many other sources. This problem is a generalized OSHA health and safety problem, and one we can have a good handle on, but it is well worth pursuing.

The reduction in cost of memory enabled by nanotechnology is also a potential risk. Memory, as we have discussed, can be used for good or for ill.

The idea that self-reproducing "nanobot assemblers" could get out of control is just nonsense. Some people imagine futurist nanoscale submarines that hunt cancer cells in the blood, but this is not going to happen. Real nuclear submarines are on the order of 100 meters in length, and they contain a vast complexity of equipment. A nanoscale submarine might be 100 nanometers in length, but this would imply it would be only a few hundred atoms long, and this is simply not enough to mimic even a few of the functions in a genuine submarine. You have to work with things that are bigger to get complex function. Nature has gone through this problem in detail, and that is why living cells are ten microns across, not nanoscale.

Education and Policy Issues

Very challenging issues concerning education blend broadly into issues having to do with science and technology in the United States. At the moment, physical science and engineering are suffering in the United States, and this raises questions. Do we really wish, as a nation, to neglect the science and technology that is doing the most to change the world? Much of nanotechnology is going to be physical science and engineering, but multidisciplinary. We need people who are educated in a number of different disciplines. Specialized nanotechnology curricula may not be needed, but they could perhaps attract students who are currently attracted to the biological sciences but not to the physical sciences. Future nanotechnologists need to know how to solve problems in three dimensions, and how to make an organic compound, and how to use Maxwell's equations, and how to think about the cell, but they don't need to know very specialized nanoscience.

A second issue in education is a shift going on in the world from "knowing" to "finding." If information is readily available, there is no need to memorize it, but there is a need to learn how to locate and use it. On a disc the size of the face of a wristwatch, we potentially can put the equivalent of about 1,000 CDs, more information than most persons know and use. Information is becoming free. An entire *world* of information is available. How do people find it? Finding is becoming in some ways more important than knowing, and this fact requires a fundamental shift in the way we have to think about education.

A third issue is that our current INS policies on immigration and naturalization are potentially catastrophic for technology. The United States has relied for 50 years on superb mathematicians from Russia, on great engineers from Japan, on very good Chinese students who come to the United States. It is much harder for them to get a visa now. This has two results. One is that fewer will come to the United States, and the other is that they will work for competitors.

Intimately related to education is public understanding, which is currently confused on the subject of nanotechnology. This confusion means that there is a real obligation on the part of the scientific community to try to help "unconfuse" people. So long as everyone is confused, imagined risks will stand in the way of real progress.

A related question is how to educate the venture capital community and industry better in dealing with some of these problems, because venture capitalism has overspecialized in the biomedical area and information.

Policy makers need to confront the issue of how to balance the research investment in nanoscience and nanotechnology between applied and fundamental exploratory research and development. This balance requires defining the role of the national initiatives.

A very important issue in this area of small technology is the need to avoid over-investing in things that are of a specific dimension; "small" is what's important. The micron scale is actually much more important than the nanometer scale right now in biology for a variety of reasons, but the nanometer scale is more important than the micron scale in electronics.

Policy decisions about commercialization must involve business, of course. Like the technology, however, business comes in different sizes. Small start-up companies are important for innovation, but with venture capitalism sputtering it is uncertain how we can make science into technology now. Large businesses will play crucial roles in development, manufacturing, distribution, and globalization. There is a gap now between our superb university system and businesses that are really focused on globalization and trying to understand how to deal with the rigors of a global environment.

Maximizing the Benefit from Nanotechnology

One of the chief benefits is certainly going to be dense information. A second is certainly going to be portability, because nano-enabled information technology will be low power, lightweight, and small volume. And a third will certainly be the ability to understand what goes on inside the cell.

How does one think about who owns what information, and what should be the restrictions on the use of information? It has been estimated that in the year 2010 one could buy 15 petabits of data storage for $250,000. That is 10^{16} bits. The human genome is about 10^{11} bits. So this means that for a few millions of dollars, I can, in principle, put everybody's genome in the United States into storage. And then for a few million more dollars, I can add credit card records, telephone logs, travel histories and things of this kind. As a free and open society, how do we want to think about the use of that information? Who owns it? What are the restrictions on it? This possibility has the potential to change society radically.

A *culture of connectivity* is a second big real impact of smalltech. With lightweight and portable information free, we are all going to be connected; all of us, to everyone, all the time. You will be able to call me wherever I am, be able to know in principle where I am, what I am doing. In principle, you can know what I have done in the past and what my intrinsic genetic capabilities are.

Thus, nanotechnology has many potentially great impacts, including portable technology; perhaps new technologies like applications of quantum effects; new small technologies like microclinical equipment and diagnostic sensors; ultra-precision manufacturing. It will invigorate capitalism with high-quality jobs, for better or worse. One result may be strengthened national security in an age of asymmetric warfare and terrorism, through global surveillance and universal tactical and strategic awareness. This constitutes a revolution in military affairs, in which the whole idea is to take small groups of people and put enormous capability in their hands through very small systems. Yet this may also lead to a loss of privacy among those whose security is protected, through very large databases, quantum computation, decryption and universal genomics.

For scientists, the nanotechnology revolution will bring a new understanding of nature: complex systems, materials, biomachines, single molecules and the cell. Yet for most people the discoveries and inventions may produce alienation, because few will have the education to understand them. As Arthur C. Clarke observed, "Any sufficiently advanced technology is indistinguishable from magic" [10, p.21, note 1].

Finally, we must address the question of how we appropriately share the benefits of this new technology with countries that are less economically well off than we, in such a fashion that there is some equity to the global development, and nanotechnology does not aggravate technological segregation of the world's societies.

References

1. M. T. Björk, B. J. Ohlsson, T. Sass, A. I. Persson, C. Thelander, M. H. Magnusson, K. Deppert, L. R. Wallenberg, L. Samuelson, One-dimensional steeplechase for electrons realized, *Nano Letters* **2**, 87–89 (2002).

2. X. Duan, C. M. Lieber, Laser-assisted catalytic growth of single crystal GaN nanowires, *Journal of the American Chemical Society* **122**, 188–189 (2000).

3. D. S. Goodsell, *The Machinery of Life*, New York: Springer-Verlag (1993).

4. W. U. Huynh, J. J. Dittmer, W. C. Libby, G. L. Whiting, A. P. Alivisatos, Controlling the morphology of nanocrystal-polymer composites for solar cells, *Advanced Functional Materials* **13**, 73–79 (2003).

5. L. S. Li, J. Walda, L. Manna, A. P. Alivisatos, Semiconductor nanorod liquid crystals, *Nano Letters* **2**, 557–560 (2002).

6. L. S. Li, A. P. Alivisatos, Semiconductor nanorod liquid crystals and their assembly on a substrate, *Advanced Materials* **15**, 408–411 (2003).

7. L. E. Manna, C. Scher, A. P. Alivisatos, Shape control of colloidal semiconductor nanocrystals, *Journal of Cluster Science* **13**, 521–532 (2002).

8. W. E. Moerner, M. Orrit, Illuminating single molecules in condensed matter, *Science* **283**, 1670–1676 (1999).

9. T. W. Odom, J.L. Huang, P. Kim, C. M. Lieber, Structure and electronic properties of carbon nanotubes, *Journal of Physical Chemistry B* **104**, 2794-2809 (2000).

10. C. Clarke, Hazards of prophecy: the failure of imagination, in *Profiles of the future: An inquiry into the limits of the possible*, Revised Edition, New York: Harper & Row (1973).

3. WORKSHOP BREAKOUT SESSION REPORTS

10 RESEARCH AND POLICY THEMES

The natural scientists, social and behavioral scientists, engineers, policy makers, philosophers, and legal experts participating in the December 2003 workshop divided into 10 panels, each focused on one theme: productivity and equity; future economic scenarios; the quality of life; future social scenarios; converging technologies; national security and space exploration; ethics, governance, risk and uncertainty; public policy, legal and international aspects; interaction with the public; education and human development. Each group produced a summary, reporting the current state of knowledge; anticipated developments; appropriate research and evaluation methodologies; areas of needed research, education and infrastructure development; and action recommendations. Summaries of the conclusions from each of these 10 panels follow.

THEME 1: PRODUCTIVITY AND EQUITY

Moderators: Mihail Roco (National Science Foundation) and Marie Thursby (Georgia Tech)

Contributors: James Adams, Mark Andrews, John Belk, Jared Bernstein, William Boulton, Georg G. A. Böhm, James Canton, Ken Chung, Julia Clark, J. Bradford DeLong, Richard Freeman, Robin Hanson, Louis Hornyak, Evelyn Hu, Peter Hébert, Laurence Iannaccone, Bruce Kramer, Joseph Reed, James Rudd, Jeffrey M. Stanton, E. J. Taylor, George Thompson, Raymond K. Tsui, Sarah Turner

Introduction

Because nanotechnology enhances many other technologies and has a great variety of potential applications [1, 2, 3, 4, 5, 6], major national efforts should be focused on how nanotechnology can increase productivity in manufacturing, improve energy resources and utilization, reduce environmental impacts of industry

53

M.C. Roco and W.S. Bainbridge (eds.),
Nanotechnology: Societal Implications — Maximizing Benefits for Humanity, 53–119.
© *2007 Springer.*

and transportation, enhance healthcare, produce better pharmaceuticals, improve agriculture and food production, and expand the capabilities of information technologies (see Figure 3.1 [4]). These areas are of general societal interest and span multiple industrial areas and research disciplines. Education about nanoscience and nanotechnology that underscores the unity of nature and manmade systems at this scale must start early in the educational process.

The panelists made estimations of the rate of introduction of nanotechnology in their own companies and the sector they represent. They reached the conclusion that by 2015, a minimum of 50 percent of new products in manufacturing and medical approaches are expected to be affected by nanotechnology. Nanotechnology is already a part of industrial processes and products—both in several traditional areas of application and in emerging ones. Nanotechnology has already affected industry, basic scientific research, education, and the global economy. These impacts are expected to continue to accrue over the coming decades. One must proactively address future implications of nanotechnology for their impact on the public perceptions of science and technology, the skill and wage composition of the workforce, the economic competitiveness of U.S. industries, the readiness of students to become nanotechnology professionals, and the degree to which benefits of nanotechnology will be equitably distributed across different areas of society.

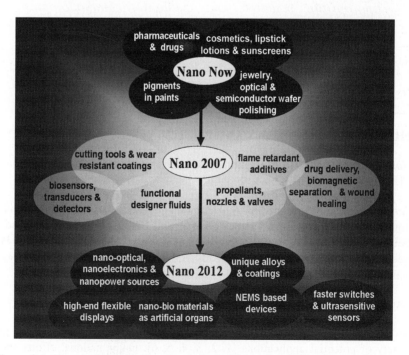

Figure 3.1. Nanotechnology timeline (adapted from *Small Wonders, Endless Frontiers*, National Research Council Review of the NNI, 2002).

Current State of Knowledge

Nanotechnology is already widely used in a variety of established industries such as catalysts, coatings, paints, rubber and tire products, microprocessor manufacturing, heavy equipment manufacturing, and aerospace. In some situations, nanotechnology is applied even when the characteristics and capabilities of the nanomaterials are not fully understood or not used at their full potential. In many cases, customers probably don't know that the products they are buying use or contain nanomaterials. Nanotechnology research and development tools are gradually accumulating, and product development processes have already begun to accelerate.

The desire for return on investment currently appears to drive incremental development and deployment of nanotechnology in existing companies and for existing types of products. For the future, so-called "disruptive changes" to manufacturing processes and product developments are expected as is the development of new industry categories and product categories. A flow of capital into new ventures has started over recent years and is increasing, but it appears that the investment process in nanotechnology needs a more coherent multi-stakeholder strategy between government, industry, universities, and venture capitalists.

Anticipated Developments

For the near future, developments in particular industry sectors are likely to be driven by the unique characteristics of particular nanotechnology processes that are immediately applicable in their respective areas. Better materials will influence manufacturing and construction; lower energy demands of devices and lighter weights of materials will influence transportation; miniaturization will influence medical diagnostics and devices and improve computing and sensing devices; and storage capacities will influence the information technology industries. These constitute non-disruptive, incremental changes that will occur gradually through reduction of manufacturing costs and improvements in product capabilities (for example, Figure 3.2).

In contrast, the disruptive effects of nanotechnology are much more difficult to anticipate because many scientific and engineering breakthroughs have yet to occur, and because nanotechnology is just one piece of a larger set of converging technologies that includes biotechnology, information technology and cognitive technologies [7]. Nanotechnology generally is not a product in and of itself, but is typically symbiotic with other technologies. Nanotechnology may serve as an *enabler* of a new product or industry category, but probably will not be that category itself. Disruptive effects of nanotechnology will probably be linked to the development of new industries and product categories (for example, Figure 3.3).

Stronger lightweight steel

Lubricant-free bearings

Real-time payload measurement

Electric muscle

Zero-emission power source

Self-cleaning surfaces

Electric Drive

Figure 3.2. Productivity enhancements envisioned for CAT heavy equipment through nanotechnology (courtesy of Caterpillar Inc.).

Healthcare

- Nano-bullets
- Artificial organs

Infrastructure

- Recyling and waste management
- Real-time toxin detection and cleanup
- Stronger structures

Consumer Electronics

- Nano-computers
- Self-assembly line
- Mass production

Energy

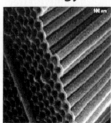

- Nano-battery
- Cleaner energy

Figure 3.3. Samples of integrative technologies from the nanoscale (courtesy of C. Montemagno, UCLA).

Research and Evaluation Methodologies

Better tools and measures for understanding the social and economic implications of nanotechnology are needed. Research methods and government funding for research in the economic, social, and behavioral sciences will allow social scientists to evaluate nanotechnology developments and provide input toward the improvement of potential applications. This will be one of the first times in history that social scientists have had such a participatory role in a technology's development. To have such a role, it is necessary for social scientists to proceed beyond the use of published, aggregate-level data in econometric studies in order to get inside the research and development processes as they occur. The techniques of interpretive social research (e.g., interviews and focus groups) as well as original data collection with surveys will enhance our ability to understand societal and economic effects much more than archival data [8]. Researchers need to talk directly to and listen to business leaders, nanotechnology researchers and nano-product development personnel. With respect to studying the possible impacts of disruptive technologies on public perceptions, new research methodologies are needed (such as "preview" respondent communities) that can provide prospective information on social impacts prior to the mass deployment of the new technology.

Research, Education and Infrastructure Development

A better understanding of the complementary influences of private, venture, university, and government investment in nanotechnology development will almost certainly improve our understanding of how best to mix these sources of capital. We also need to understand impacts on productivity, diffusion of innovations from basic science into applied engineering, rates of commercialization, and U.S. industrial competitiveness internationally. We need systems-oriented research to examine the interlocking processes of research, development, and technology diffusion.

A stream of research is needed that specifically examines equity issues. Elements of this research could be aimed at understanding technology development impacts on underserved populations. Other research should examine how market and non-market forces affect distribution of benefits expected from nanotechnology. Cross-national research is needed that examines economic effects of nanotechnology in less developed nations. Research could examine the distributions of risks and benefits across different segments of the population.

Besides the specific topic of equity, a considerable body of new social impacts research is needed at the individual, group, and societal level. We need to understand and predict changes to workforce composition, skill-biased technological change, and public reactions to new technologies, among other areas.

In addition, we need research and infrastructure that can ask how K-12 science and technology education can be enhanced to address science and engineering opportunities for underprivileged populations. This will permit universities to build upon a more sophisticated base of knowledge than is currently available. An apparent educational need is for individuals who possess both a depth of knowledge and greater abilities to communicate across disciplines. Individuals who can communicate across two or more fields, e.g., physics, chemistry, biology, information science, engineering, management, social sciences, law, and humanities are particularly valuable. Attention should focus on educating for a variety of roles, not just educating future scientists.

Action Recommendations

A programmatic approach is needed to increase synergy in nanotechnology development by creating partnerships earlier in R&D processes between industry, academia, national laboratories, and funding agencies. Working to develop an understanding of societal impacts before and during the creation of new technologies will help generate understanding of co-evolving developments in science, education, and production. Deliberate actions are needed for better equity in distribution of the potential benefits of nanotechnology. Several strategies are recommended:

- Multi-functional clusters or partnership coalitions should be created that bring together those involved in researching and developing nanotechnology and other technological tools, processes, and products.

- Government and the private sector should anticipate impacts to the extent possible and mitigate negative impacts on people, such as workers in obsolete industries, whose lives may be disrupted by the nanotechnology-related advanced.

- Measures should be taken toward balanced distribution of benefits and risks in society, between public and industry advantages.

- Workforce development should be undertaken across the full spectrum of job roles (i.e., not just scientists), for example combining traditional educational strategies with internships, retraining programs and continuing education.

- Increased capabilities and funding should be developed for conducting science and technology studies in educational contexts, in industrial contexts and among the public.

References

1. NSTC, *National Nanotechnology Initiative: The Initiative and Its Implementation Plan*, Washington, D.C: NSTC (2000) (available at http://www.nano.gov/html/res/nni2.pdf).

2. NSTC, *National Nanotechnology Initiative: Supplement to the President's FY 2004 Budget*, Washington, D.C.: NSTC (2003) (available at http://nano.gov/html/res/fy04-pdf/fy04-main.html).

3. Chemical Industry Vision2020 Technology Partnership, *Chemical Industry R&D Roadmap for Nanomaterials by Design: From Fundamental to Function*, Chemical Industry Vision2020 Technology Partnership (2003) (available at http://www.chemicalvision2020.org/nanomaterialsroadmap.html).

4. National Research Council, *Small Wonders, Endless Frontiers, A Review of the National Nanotechnology Initiative*, Washington, D.C. (2002).

5. Semiconductor Industry Association, *The International Technology Roadmap for Semiconductors, 2003 edition*, Austin TX, International SEMATECH: (2003).

6. M. C. Roco, R. S. Williams, P. Alivisatos, eds., *Nanotechnology Research Directions*, Washington, D.C.: U.S. National Science and Technology Council (1999), also Dordrecht: Springer (formerly Kluwer) (2000) (available at http://www.wtec.org/loyola/nano/IWGN.Research.Directions).

7. M. C. Roco, W. S. Bainbridge, eds., *Converging Technologies for Improving Human Performance: Nanotechnology, Biotechnology, Information Technology and Cognitive Science*, NSF-DOC Report, Washington, D.C.: NSF (2002), also Dordrecht: Springer (formerly Kluwer) (2003).

8. M. C. Roco, W. S. Bainbridge, eds., *Societal Implications of Nanoscience and Nanotechnology*: National Science Foundation Report, Arlington, VA: National Science Foundation (2000), also Dordrecht: Springer (formerly Kluwer) (2001) (available at http://www.wtec.org/loyola/nano/NSET.Societal.Implications/).

THEME 2: FUTURE ECONOMIC SCENARIOS

Moderators: Michael R. Darby (University of California, Los Angeles and National Bureau of Economic Research) and Daniel H. Newlon (National Science Foundation)

Contributors: Ilesanmi Adesida, Mark J. Andrews, Hongda Chen, J. Bradford DeLong, Richard Freeman, Robin Hanson, Geoff Holdridge, Louis Hornyak, Peter Hébert, Thomas A. Kalil, Scott McNeil, David Mowery, Sean Murdock, Linda Parker, and Mihail Roco

Introduction

Breakthroughs in nanoscale science and engineering are expected to have potentially profound effects for future economic conditions. This process of accelerated change and metamorphic economic progress—already underway—is uncertain as to specific impacts and products. Metamorphic progress is relatively rare, usually occurring in only a few industries at any time; over time, however, it has marked great advances in our standard of living, such as spinning, weaving, steam engines, steel, glass, automobiles, aircraft, electricity, semiconductors, information technology, and biotechnology [1, 2, 3].

Current State of Knowledge

At present, there is good reason to expect that nanotechnology will impact a wide range of industries—simply lowering input costs in some industries, dramatically improving productivity in others, creating entirely new industries, increasing demand for some goods and lowering demand for others. This is characteristic of major innovations in a dynamic economy, with innovators and their emulators, both established and new firms, lowering the cost of living for consumers even as they displace producers who can no longer produce the products as cheaply. While society as a whole benefits, owners of specialized resources—be they equipment, natural resources, or industry-specific labor skills—may be hurt or helped. These risks of ownership are viewed as an unavoidable cost of living in a dynamic, capitalist economy in which each generation can expect its children to have higher standards of living than its own.

A few examples may help. The invention of the automobile led to many start-up manufacturers as well as transforming former buggy manufacturers. Other buggy manufacturers who could not make the change successfully were gradually forced out of business in the face of declining demand, as were most stables and the proverbial buggy-whip makers. Costs were lowered for transportation and delivery firms, and those industries grew as their customers demanded more of their services at lower prices. Petroleum demand increased dramatically. There were winners and losers in the short run, but society as a whole clearly benefited. Henry Ford's invention of the automotive assembly line dramatically reduced the cost of automobiles to consumers and increased the number sold, but relatively few successful emulators were able to compete and prosper in the dramatically expanded market, and hundreds of small auto producers were forced out of business in the face of mounting losses.

The market system ensures that major, disruptive innovations cause only a transitional increase in unemployment as labor and capital are shifted to new, more valuable uses from those that have been superseded or made less valuable. Real wages and the standard of living are closely tied to growth in labor productivity resulting from both increasing levels of education and from innovations. We know of no case in which negative transitional effects of innovations were of such significance that society would have been better off if some of the innovation—and consequent increases in average wages and standards of living—had been postponed.

Anticipated Developments

Over the long term, beyond the next decade, many major economic impacts are expected from nanotechnology. Major impacts will occur in the semiconductor industry, allowing continued progress in computer chips and probably in sensors. It is possible that we will see the initial successful products from the convergence of nanotechnology with biotechnology. Other anticipated developments in the 10-20 year time scale are a revolution in production of many materials, reducing cost and improving performance, and the first applications of nanotechnology to pollution remediation, which reduce the cost of cleaning up negative impacts of traditional technologies, such as those covered by the Superfund. This panel expressed the opinion that over the next 25 years, but probably not as soon as the next decade, we would expect to have the technical means to produce significant amounts of energy by a host of conversion solutions and improved efficiencies ultimately attributable to nanotechnology. In the long term, there are also likely to be substantial health benefits, including major decreases in morbidity and mortality, helping people live better and longer.

Predicting the course of emergent technological trajectories—much less their economic impacts—is always challenging. Wide allowance must be made for inevitable error. In addressing future economic effects, researchers need to take a number of different but ultimately complementary viewpoints. One point of view addresses the macroeconomic effects in terms of effects on economic growth, productivity, real wages, and the standard of living. Another industrial organization approach focuses on the particular industries that will be most directly affected by nanotechnology and attempts to assess how they will be transformed. Labor economists take a similar point of departure, but instead focus on the movement of labor among industries, the acquisition by some workers of new skills, the obsolescence of other skills, and the necessary wage signals to accomplish both labor movement and training.

For each of these viewpoints, the first step is framing the "counter-factual question"— what is the effect of the nanotechnology that we are going to see compared to what alternative? One conceivable alternative is that scientists never learned to work on the nanoscale and accordingly spent their time doing something else. But this alternative is intractably abstract, so it is worth considering other more tractable and meaningful counterfactual situations to serve as a basis for comparison. For example, one could compare the situation if nanotechnology were used extensively in the United States but not elsewhere, or whether the government will or will not take a leading role in investing in nanotechnology development.

The economic effect of using nanotechnology—or hypothetically banning its use—depends on whether the United States is alone in using, or in not using, the technology. Indeed, there is an argument to be made that the economic benefit of an international ban on use of nanotechnology would depend on whether the ban was across-the-board or concentrated only in particular regions, creating competitive disadvantages for those regions. Another interesting problem concerns the economic return to government investment in nanoscale research and if and how that is impacted by foreign governments making similar investments.

Industrial organizations and labor economists will find it difficult to specify the nature and magnitude of impacts on various industries. We believe those assessments would be better if the National Nanotechnology Initiative initiated ongoing collection of data that detailed the use of nanotechnology by industry and geography. These data could be collected and reported annually either by the Economic and Statistics Administration of the U.S. Department of Commerce or by private or academic contractors. It is hard to predict where we are going without knowing where we already are.

Macroeconomic assessments of technological change have become standard practice and the approach is reasonably well framed. While nearly all goods and services would be impacted to some extent by improved energy efficiency and improvements in materials and computing power associated with nanotechnology, macroeconomists would not say that therefore nearly 100 percent of output will result from nanotechnology. Rather the approach is to measure the increase in the real value of goods produced from a given quantity of factors of production. This increase—often quoted as percentage of GDP—is the ultimate measure of how much the standard of living is increased by nanotechnology in one specific country. Given our experience with other technologies of broad application, we anticipate that future research will be able to document impressive social benefits from investment by governments in nanotechnology development.

Research and Evaluation Methodologies

Economists should be encouraged to address six main research questions, in support of the National Nanotechnology Initiative.

First, what policies increase or deter the transfer of nanoscale science and engineering knowledge from academe to industry? Issues to be examined include licensing intellectual property to inventor-affiliated companies under the Bayh-Dole Act and concerns about conflict of interest or commitment. The issues could be framed in terms of the optimal assignment of property rights for university research, developing ideas of Aghion and Tirole [1] and Jensen and Thursby [4].

Second, what are the returns from nanotechnology investment? This question addresses why the government should be investing in nanotechnology. An interesting alternative question is: How much should government and industry invest, given a reference rate of return?

A third research question, with both economic and social dimensions, concerns how longer, healthier lives will change work patterns. If people are able to work far into their seventies, will they and the labor market around them become segmented by age? How will the activity of older people be distributed across entertainment, paid labor and volunteer contributions to the well-being of others?

Fourth, what skill biases are associated with major nanotechnology applications and what do they imply for wages and returns to education?

Fifth, should there be a research exemption for patents? At present, there is much debate about the conditions under which scientists should have a free license to employ patented inventions in non-profit research, for example, in their instrumentation and other research tools. Traditionally, technological inventions can be patented, whereas scientific discoveries cannot. Yet, the line between nanotechnology and nanoscience is unclear, and the economic benefits of progress might be diminished if intellectual property rights prevent rather than stimulate innovation.

Finally, what are the most efficient and effective forms of government-industry-academe research cooperation? NSF's Grant Opportunities for Academic Liaison with Industry (GOALI) and the Microelectronics Advanced Research Corporation/ Department of Defense Focus Center Research Program are examples that deserve consideration.

The research methods that are most likely to result in usable, credible answers to these questions are: econometric analysis; major data collection (firms, products, people); Nanobank.org and studies of NNI grant awardees; Center for Economic Studies/ Research Data Series; case studies (e.g., industry, cross-cutting tech); surveys (e.g., technology transfer, job skills, Delphi); economic models (input-output models, dynamic adjustment); and input-output analysis of impact on sectoral productivity.

Research, Education and Infrastructure Development

A nanotechnological revolution would have implications for education and infrastructure. The first key point is that top scholars and doctoral students are the scarce resource around which industries will be formed and transformed. It is therefore in the national interest that we do all that we can to encourage top foreign talent to come to the United States and work here rather than building foreign capabilities.

Business schools should offer courses to familiarize both MBA students and mid-career executives with the basic ideas and the potential of nanotechnology. Conversely, there is an important niche in facilitating the aspirations of scientists attempting to become or work with entrepreneurs in order to bring their discoveries to market.

The research questions listed above document the important role that economic research can play in resolving major issues in how much government should invest in nanoscale research and how to make those investments have the greatest return to the national welfare.

Finally, this revolution is one more reason to stress the long-term importance of strengthening K-16 math and science education. Applications to entrepreneurial activity can provide a powerful hook for those students who are not intrinsically motivated to study math and science for its own sake.

Action Recommendations

The accumulated research to date makes a strong and clear case for one recommendation that is most important for maximizing the benefits from investment in nanoscale science and engineering: research should be broadly funded, primarily based on peer-reviewed investigator-initiated proposals; benefits of nanoscale science and engineering are so broad that funding should not be driven by a few specific top-down priorities.

References

1. P. Aghion, J. Tirole, The management of innovation, *The Quarterly Journal of Economics* **109**, 1185–1209 (1994).

2. M. Darby, L. G. Zucker, Growing by leaps and inches: Creative destruction, real cost reduction, and inching up, *Economic Inquiry* **41**, 1–19 (2003).

3. C. Harberger, A vision of the growth process, *American Economic Review* **88**, 1-32 (1998).

4. R. Jensen, M. Thursby, Proofs and prototypes for sale: The tale of university licensing, *American Economic Review* **91**, 240–259 (2001).

5. M. L. Tushman, P. Anderson, Technological discontinuities and organizational environments, *Administrative Science Quarterly* **31**, 439–465 (1986).

THEME 3: THE QUALITY OF LIFE

Moderators: Michael Heller (University of California, San Diego) and William Bainbridge (National Science Foundation)

Contributors: Robert Beoge, Elaine Bernard, Rosalyn Berne, Nila Bhakuni, Stan Brown, Tanwin Chang, Hongda Chen, David A. Diehl, Toby Ten Eyck, Michael

Heller, Stephan Herrera, Barbara Karn, Kristan Kulinowski, Richard Livingston, Donald Marlowe, Carlo Montemagno, Sean Murdock, Günter Oberdörster, Steven Papermaster, Cheryl L. Sabourin, Jeffery Schloss, Richard Smith

Introduction

Rapid progress in nanoscience and nanotechnology offers a positive vision of the future. The National Nanotechnology Initiative will proactively fund R&D for new nanoscale capabilities to ensure the maximum improvement of the quality of life at both the individual and societal levels. At the global, as well as local levels, we must act wisely to improve the sustainability of the world around us. Four key areas are food, water, energy, and preservation of the environment.

Nanotechnology will help ensure that we can produce enough *food*. Importantly, it will also improve shelf life, packaging, the ability to grow locally, labeling, product identity, and history tracking. Nanotechnology will also help with *water* resources, allowing low-energy local purification (chemical and biological) and desalination, while also reducing water waste in manufacturing and farming. Nanoscale-related improvements in *energy* technology will reduce our dependence on fossil fuels, make photovoltaic energy production competitive with other sources, facilitate entrance into a hydrogen economy, and improve renewable energy systems like biomass. In order to preserve the *environment*, we must use nanotechnology to remediate air and water pollution, produce systems and materials that contribute to reducing resource consumption and waste production, facilitate waste remediation, recycle pollution into raw materials, and ensure safety and sustainability of new materials.

Over the next 20 years, nanotechnology will likely contribute greatly to improved human health and well-being, as better and more affordable medical diagnostics and treatments lead to increased longevity at comfortable, active, and productive levels of vitality. Another vast area of benefit will be improved detection of hazardous contaminants, such as nuclear, biological, and chemical agents in the food chain and in population centers. Nanotechnology-enabled devices and nanoscale-structured materials will increase the mobility, safety, and efficiency of transportation systems. In their everyday lives, people will benefit from greatly enhanced tools for cognition and communication. Participants in the breakout group hypothesized that by increasing the total wealth at the disposal of humanity and by transforming inefficient delivery systems into efficient ones, it is conceivable that nanotechnology could achieve a better distribution of economic benefits throughout the population.

Current State of Knowledge

Scientists and policy makers can dimly foresee potential manufacturing and process systems in many spheres, but they cannot predict how processes that do not yet exist will be implemented nor what the indirect effects and associated societal implications might be. Much basic research is needed into the ways nanomaterials and nanosystems intersect with biological and ecological systems. Because nanotechnology is so broad-based, with potential evolutionary and revolutionary impacts, it may be easy to understand implications in one case, but not in another. Numerous paradigm changes may be required. It is clear that we have little understanding of the system-level effects, such as unintended consequences, that derive from the complexity of nanotechnology. If societal leaders have such great difficulty seeing into the future, it is not surprising that the general public lacks a clear understanding of risks and benefits to the individual, of the balance between advantages and disadvantages, and of how to distinguish fanciful hype from pragmatic reality.

Anticipated Developments

During the next decade or so, small, fast, and dumb nanodevices will control the next generation of computers and extend Moore's Law in computing. Old but never fully developed concepts such as artificial intelligence and virtual reality will become realizable, creating advances in computing, entertainment, and human interaction. Some participants also expressed concerns about the society's capacity to adjust well when fast rates of change in science and technology occur. Diseases will be diagnosed and treated far more rapidly, allowing people to live years longer.

As nanoscience gives birth to nanotechnology applications, we can expect every sphere of human life to be affected in one way or another. New materials will replace older materials and, in so doing, may render obsolete the industries and industrial bases that had been built around the latter. Governments will face the policy decision of whether to let obsolete industries die quickly or to subsidize them for a time to reduce potentially harsh impacts on employees.

As new nanoscale-related technologies become available, some people will have greater access to them than other people, depending on such variables as social class and level of national development. How much inequality there should be in society is partly a political question, but clearly, the general economy benefits from increasing the fraction of the population who are prosperous consumers. Privacy could become increasingly rare as cameras and microphones grow smaller and smaller. Policy questions abound, but few answers are apparent. The fields of medicine and biotechnology are already fraught with ethical dilemmas, and the convergence of technologies may complicate a number of extant policy issues.

Research and Evaluation Methodologies

Forecasting, scenario-building and other future research tools will help tease out the possible landscapes of the world to come. Opinion polls and surveys will help tell us where the public currently stands on nanotechnology-related issues, and the full range of social science methodologies will be helpful. Facilitated seminars will provide public education while allowing the compilation of a history and constituting a natural source of survey respondents, so that the changing range of attitudes can be assessed over time. Each research method has its place, but it will be impossible to rely upon any one alone. Rather, it will be necessary to take a systems-oriented approach, studying not only each part of the socio-technological complex that constitutes civilization, but also examining how it fits into the larger dynamic system.

Important lessons can be drawn from the Ethical, Legal, and Social Implications (ELSI) program associated with the Human Genome Project, lessons that are relevant to a nascent program in Societal Implications of Nanotechnology. A key focus of the ELSI program is to support research whose results will inform development of well-founded policy. At times, the research program may have to adjust to public perceptions. An example is the development of the SNP (Single Nucleotide Polymorphisms) resource, a collection of human DNA samples that would be mined to create a public database of common variation in DNA sequence in the human population. To create a comprehensive dataset that is widely relevant to humanity, samples from wide geographic origins were needed. Engaging those communities required sensitivities to many concerns. To achieve a very desirable scientific goal, information was sacrificed even though it would be needed later for follow-up studies of disease, namely, the labeling of the geographic origin of those samples. Instead, all of the samples were collected, randomized, and deposited anonymously. The SNP resource has been widely used to generate essential data. The relationships that have been built with various communities since that time have enabled the subsequent collection of identified samples, for development of the population-specific haplotype maps. However, the decision to avoid geographic labeling of the samples retarded some scientific research and effectively excluded some scientists, notably physical anthropologists. Thus, including the public in scientific decisions offers a mixture of costs and benefits that must be carefully weighed and responsibly managed.

Any major research program will have to begin by providing as many educational opportunities as possible, not just for future nanotechnology workers but for the general public. A lesson can be learned from the debacle in Europe when agribusiness went ahead with genetically modified organisms, but without proper public support. Public participation in the discourse is essential. Even if the best and brightest scientists think they already know the answers, a policy of evidence-based argument would be far superior to one of heartfelt assertion.

Research, Education and Infrastructure Development

The key areas of research on the societal implications of nanotechnology for the quality of life include the identification of public concerns, perceptions of risk, fears, conceptions, and misconceptions. Research should identify the qualities of work, life and the environment to which citizens give highest priority, and identify the branches of nanotechnology most relevant to them. It will also be necessary to evaluate the claims made by public advocacy groups, understand how these groups came to their particular advocacy positions, and determine how they differ from the views of scientists and the wider public. Longitudinal research should chart how the competing perspectives evolve, especially to the extent that they diverge from the consensus among scientists and engineers about the implications of nanotechnology.

Other important research goals are more immediately connected with the need to frame policy over time, as conditions change and new issues arise. A means to monitor for early signs of negative aspects and risks should be developed which will permit the timely development of contingency plans to handle problems. A potential risk of great political and ethical significance is that benefits deriving from nanoscience could widen the gap between the haves and the have-nots rather than being equitably distributed as hypothesized earlier. Research should seek to understand the conditions under which this may happen, the factors that can maximize the distribution of nanotechnology's benefits throughout the population, and the adjustment mechanisms (such as social movements and government safety nets) that may best absorb any shocks that such inequality might create. Additionally, research should seek to understand what kinds of institutions are best able to safeguard the public in areas such as privacy and nanotechnology-related hazards, without inhibiting development of beneficial applications.

Action Recommendations

The following actions and anticipatory measures should be undertaken to clarify the public perspective on nanotechnology, so that optimal decisions can be made to employ this diversity of new means to improve the quality of life:

- Public education should inform the citizenry about the diversity of methods, principles, and materials that constitute real nanotechnology, as distinct from the often simplistic myths spread by some mass media and special interest groups.

- Scientifically reliable and publicly respected organizations should clearly articulate the near- term and the long-term benefits (and risks) of nanotechnology, to solidify public trust and empower people to make good nanotechnology investment decisions.

- Scientists and engineers should be forthright in stating what their research will and will not produce; a guiding vision is important, but unrealistic claims are not helpful for real progress.

- Scientists, educators, the mass media, and policy makers should clearly distinguish the direct effects of nanotechnology on the quality of life from the effects of other technologies (such as genetic engineering) that are sometimes connected to nanotechnology, but are really quite separate phenomena that require separate treatment.

References

1. National Research Council, *Small Wonders, Endless Frontiers, A Review of the National Nanotechnology Initiative*, Washington, D.C., National Research Council (2002).

2. Special supplement to *The Journal of Law, Medicine & Ethics* **29**(2) (2001).

3. Presentation abstracts for A Decade of ELSI Research: A celebration of the first 10 years of the Ethical, Legal, and Social Implications (ELSI) programs, conference held January 16-18 at the Natcher Conference Center, National Institutes of Health (2001).

THEME 4. FUTURE SOCIAL SCENARIOS

Moderators: William Bainbridge (National Science Foundation) and Roger Kasperson (Clark University)

Contributors: John Belk, Rosalyn Berne, Jeffery Schloss, James Canton, Arthur Caplan, Stephan Herrera, Laurence Iannaccone, Frank Laird, Richard Livingston, John Miller, Günter Oberdörster, Bruce Seely, Jeff Stanton, John T. Trumpbour

Introduction

Scenario analysis, as practiced here, can help identify issues and hypotheses, and thus, is a useful tool of theoretical analysis. This panel puts forth two very different scenarios for the coming 10 to 20 years, in order to help clarify both the issues related to nanotechnology that policy makers will face, and the knowledge that needs to be gained through research. Both scenarios concern the transition from the relatively crude technologies that society depends upon today to more efficient, productive, and environmentally friendly nanotechnology-enabled technologies. In one scenario, the transition will be smooth and benign, whereas in the other scenario the transition will be rough and marked by many different kinds of harm and conflicts with social values and institutions.

Smooth transition: In this scenario, nanoscience and applications based upon it will be developed in a way that allows scientists, engineers, and policy experts to deal with the unexpected. There are bound to be surprises, because every major technology has unintended consequences (which may be either positive or negative). The question is whether our institutions will be prepared to take best advantage of the positive consequences and to reduce the human or economic costs of any negative consequences that arise.

In this optimistic scenario nanotechnology produces clear, demonstrable benefits and solutions for real-world problems, and management strategies for threats. For example, it will enable low-cost energy production with minimal impact on the environment, as well as achieving greater efficiency in energy use. It will help prevent and cure disease, and will provide many rewarding jobs. It will contribute to applications that strengthen the nation's defense capabilities without unduly burdening the privacy of citizens, while also reducing the incidence of terrorist activity and strengthening the cause of peace worldwide. In this scenario, early applications stress positive effects on publicly valued areas, such as health, energy and food development, pollution abatement, and environmental protection.

Importantly, the smooth transition scenario assumes that nanotechnology development will benefit from strong public involvement. There will be a consensus about a list of uses that the public would concur are indeed beneficial. Policy makers will have access to the knowledge and explanatory tools that will allow them to evaluate accurately the relative benefits and risks of specific technologies, on a structural level of the entire society as well as locally within the application itself. The influential existing institutions of society will vet proposals in a balanced way, supporting the adoption of nanotechnology in most areas. At the same time, there will be smooth transitions involving multiple changes in the social fabric that contribute to multiple positive future scenarios.

Rough Transition: This scenario could lead eventually to a happy situation like that described in the smooth scenario, but only after a longer period of delay and with very substantial human costs. Clearly unmanaged or unanticipated risks become evident with this scenario. At the extreme, it could lead to the near-permanent abandonment of some forms of nanotechnology and thus to a failure to take advantage of their benefits.

Societal institutions and the general public would not be effectively involved in the policy-setting process. When surprises come along, the leadership and affected people will be unprepared to deal with them. In some cases, heavy, but poorly planned, investment in a particular branch of nanotechnology might lead to profound disappointment. Perhaps nanotechnology-enabled weaponry would be used in such a way as to increase rather than decrease fatalities, ultimately leading to reduced security. The public would perceive that industry and scientists are concerned only with their own profits and careers, causing widespread apprehension and mistrust. There could be irrational fads leading to government regulation that was either too rigid or too lax, and a tremendous loss of investment coupled with tragic failures to realize the greatest benefits of nanotechnology.

There are parallels in previous technology revolutions or evolutions that can inform us about the future of nanotechnology. Genetically modified organisms, stem cells, and nuclear power exemplify the rough transition scenario. Research is needed to compare nanotechnology with the history of these other technologies to draw lessons that may be relevant. This must be done in full awareness of the fact that nanotechnology is a highly varied collection of different scientific and technical capabilities, with recognition that most of them may prove to be quite different from past technologies, both in actual societal impact and in the ways that people respond. Indeed, one of the more disturbing possibilities is that policy makers and leaders of social movements may respond to nanotechnology not as it actually is, but in terms of false analogies.

Current State of Knowledge

At present, it is difficult to distinguish between those impacts that are distinctive to nanotechnology and those arising from the synergistic effects of the convergence among technologies. In this convergence of nanotechnology with other technologies, nanotechnology could potentially be contaminated by controversies or stigmas that already exist for those other technologies. The most obvious current example is the connection between nanotechnology and genetic engineering. Much of actual nanotechnology involves inorganic chemistry, materials, and physical processes that are quite remote from those of biology. However, a few leading researchers are explicitly exploring how nanoscale-engineered components might be integrated with living cells. Beyond these real experiments, both science fiction authors and various social movement organizations have envisaged possible intimate nano-bio interrelations that seem to have captured the popular imagination. Such issues must be separated from considerations of the societal impact of the great majority of nanotechnology applications.

We do not know the extent to which the dissociation of risks from benefits as a result of nanotechnology applications will produce significant inequity problems among groups, countries, places, and generations. Another way to put this point is to say that social science cannot immediately say how the new complexity of nanotechnology will interact with the existing complexity of society for the benefit of various groups.

Anticipated Developments

There is the potential that some applications of nanotechnology will be expensive, creating "haves" and "have nots" for some products or services. The "digital divide" of information technology, for instance, affected some groups of people. Some people are concerned that inequalities could be created, for example, between developing and developed countries or between classes within a society. The industrial revolution, beginning notably in Britain over two centuries ago, led to increased wealth and eventually to a healthier life for most people, but the short-term impact on workers may possibly have been negative, both because of pollution and job-related hazards and because existing social institutions had a limited capacity to deal with the new channels of power in the society. If nanotechnology causes sudden changes in production and wealth, and if societal institutions exhibit cultural lag, then initially some categories of people may suffer as others benefit.

71

On the other hand, gradual implementation, as will happen as vast numbers of nanotechnology-enabled improvements are introduced across many industries, could benefit everyone. If some people benefit more than others, then it is a political decision as to whether policies are needed to equalize the benefit to some extent.

Nanotechnology will contribute to general economic growth, and growth provides the resources to deal with social problems more effectively. However, it would be utopian to believe that any technology alone can solve major social problems. For example, the direct effects of nanotechnology on world famine and hunger are not likely to be large, as history shows that these conditions are caused primarily by policy, war, and distribution problems.

The effect of nanotechnology on higher education could be quite significant, in conjunction with the new financial realities faced by universities. Currently, university scientists face much pressure to launch start-up companies and to collaborate more with industry. It is controversial whether this is a good thing. Some people argue the value of keeping nanotechnology in the public domain, but it is not clear how this may be done without impeding the development of the technology.

Conversely, many academic fields and individual scientists or engineers may benefit intellectually from improved connections to commerce and industry. Appropriate policies, at the national as well as campus level, can ensure that scientist-industry collaborations lead to more knowledge (more life-long learning) and knowledge networks or knowledge banks, not less. Nanoscience can transform global supply chains, and nanotechnology applications can be used to promote more industrial development that will in turn produce more productive and equitable trade.

Research and Evaluation Methodologies

Scenario analysis, as mentioned earlier, can help identify issues and hypotheses, and thus is a useful tool of theoretical analysis. A worthwhile variant of scenario analysis is backcasting, the mirror image of forecasting, which specifies an outcome and tries to identify the steps that might lead to it. Scenarios are an art form, akin to brainstorming, but there are ways to render them more rigorous. For example, acknowledged experts can be asked to write the scenarios, and their output can be harmonized with known facts (such as demographic data or statistics on availability of natural resources). Even when they are not fully rigorous, scenarios can help policy makers and ordinary citizens alike to imagine possible futures, both to prepare responses to anticipated problems and to set goals for positive accomplishments. Ideas generated through scenarios can become the focus of more rigorous methods of empirical research.

Multi-agent modeling is akin to scenarios, but is carried out through computer simulation. An agent is a dynamic computer representation of an individual person, organization (such as a corporation), sector of the economy, or other social unit.

Among the most intellectually influential examples is a study by political scientist Robert Axelrod [1], in which a computer modeled the interaction of a number of individual people, who followed various strategies in their economic dealings with each other. The point of the study was to see if these agents could learn to cooperate, despite the fact that each was programmed to seek his or her own best selfish interests, and indeed they could. The relevance to real people was that the study showed that cooperation between humans was logically possible even without shared social values, religion, or any of the other sophisticated cultural factors that are often assumed to help humans be reliable partners. For 30 years, agent-based and other computer simulations have contributed to a greater understanding of issues, such as the ways a society may affect the natural environment and the ways social movements may organize around a variety of issues [2, 4, 6]. Axelrod's simulations employed game theory; it is also possible to use pure mathematical methods in this way of conceptualizing human relations in terms of strategic interactions for personal gain.

The case study method is an important qualitative research approach that can be practiced somewhat rigorously, either with historical or ethnographic data. Given that nanotechnology is quite recent, historical studies will have to rely upon carefully drawn analogies with earlier technologies. For example, an extensive literature already exists on the often-rocky adoption of new medical technologies, as some excellent therapies and diagnostic tools are ignored while others spread rapidly throughout the medical community despite lack of evidence for their value [3, 5]. The challenge is how to identify close analogies between past cases and particular nanotechnology applications. The ethnographic approach avoids this problem through direct observation of a specific emerging nanotechnology in the laboratory or in the wider organization of which it is a part. Ethnography is not well suited for prognostication, however, because it focuses on the present and very recent past. Potentially, the combination of history (to get the time perspective that reveals outcomes) and ethnography (to determine the nature of an innovation to support appropriate analogies) could be more powerful than either alone.

New technologies do not merely have an *impact upon* society. Rather, they *interact with* society, and their impact is a result of technical facts with social factors. Thus, public opinion surveys and methods like market testing are important ways to chart the changing meaning of nanotechnology. Focus groups can provide insights into how to intervene and how to get information across. The research method must be tailored to the particular population under study. For example, young people may not respond well to formalized questionnaires, so it may be best to conduct listening tours in high schools to examine youth culture and understanding. Content analysis (obtained by looking at media, popular culture, and Hollywood) could be integrated with surveys of audiences for analysis of the social values that nanotechnology may affect.

Finally, it will be important to collect solid facts about the institutions and individuals that are most involved in the development and application of nanotechnology. An inventory should be undertaken of existing institutions and assessment of how they cope with change and uncertainty. Research to develop a future nanotechnology-skills inventory for identifying best-of-breed competencies that will enable jobs, career development and competitiveness would also be valuable.

Research, Education, and Infrastructure Development

The most fruitful lines of research to develop realistic social scenarios about the future value of nanotechnology for society include those that focus on the societal institutions and on formal organizations that both create nanotechnology and respond most directly to its effects. Which organizations are already attempting to deal with future scenario questions, and with what results? How adequate are the existing institutions, agencies, legislation, and rules in terms of how nanotechnology fits into existing systems? What institutional forms and processes will lead to the best balance of innovation, security, equity, and health and environmental protection?

Also important will be research on the processes of innovation, diffusion, and adjustment. What flows of talent (college, private) across national boundaries are beginning that may accelerate nanotechnology innovations? How can we better anticipate the impact of nanotechnology on institutions and society so that we may understand and manage the change process to enable productive outcomes for jobs, economy, industry, education, and human well-being? What are the major societal trends already in progress that might be accelerated or retarded by different developments related to nanoscience and nanotechnology? What social scenarios might cause or occur from specific events, such as the environmental release of nanoparticles?

Action Recommendations

To take advantage of the new technology in a timely and responsible way, the following should be considered:

- A program should be launched to open communication channels linking science and industry with religious, cultural, and moral leaders—sooner, rather than later—supporting effective ethical and economic discussions.

- Industry, government, and academia should cooperate in a campaign to assemble and disseminate timely, accurate information about nanotechnology and its social implications, using such methods as national and regional forums that bring together different groups, conferences and information networks, school curriculum, and media kits.

- Policy leaders should develop a national nanotechnology competitiveness plan, involving stakeholders interested in competitiveness. Plans might include, for example, establishing a program that enables nanotechnology entrepreneurs.

References

1. R. Axelrod, *The Evolution of Cooperation*, New York: Basic Books (1984).

2. W. S. Bainbridge, *Sociology Laboratory*, Belmont, CA: Wadsworth (1987).

3. J. P. Bunker, B. A. Barnes, F. Mosteller, *Costs, Risk, and Benefits of Surgery*, New York: Oxford University Press (1977).

4. J. W. Forrester, *World Dynamics*, Cambridge, MA: Wright-Allen Press (1971).

5. J. D. Howell, *Technology in the Hospital: Transforming Patient Care in the Early Twentieth Century*, Baltimore: Johns Hopkins University Press (1995).

6. D. H. Meadows, D. L. Meadows, J. Randers, W. W. Behrens III, *The Limits to Growth*, New York: Universe Books (1974).

THEME 5: CONVERGING TECHNOLOGIES

Moderators: John Sargent (Technology Administration, Department of Commerce) and Lynne Zucker (University of California, Los Angeles)

Contributors: Ilesanmi Adesida, James R. von Ehr II, Roger Kasperson, Judith Klein-Seetharaman, Sharon Levin, Sonia E. Miller, Cyrus Mody, David Rejeski, Nora Savage, Paula Stephan

Introduction

Over the coming decades, the world may be transformed by the convergence of four major realms of discovery and invention: nanotechnology, biotechnology, information technology and new technologies based in the cognitive sciences. This convergence is sometimes identified by the initials of the four realms, or NBIC. In recent years, it has become clear that profound potential is at the intersection points of these four technologies—in pairs (e.g., nano-bio), trios (e.g., nano-bio-IT) or altogether (e.g., nano-bio-IT-cogno). In this section, we consider the convergence of nanotechnology with one or more of the other three areas.

Current State of Knowledge

Each of the four technologies offers extraordinary potential for economic growth, job creation, national defense, homeland security, and improvements in a variety of other areas. Together, their impact is expected to be significant, as well. Because of the potential they offer, the United States Government, the private sector and governments around the world are investing billions of dollars annually in research to promote their development and enable their commercialization.

As a result of these growing investments, cumulative knowledge, and competitive forces, the pace of scientific discovery and technological progress is accelerating. Achievements that long have been dreamed of—successfully treating fatal diseases such as AIDS and cancer; improving crop yields and nutritional value to feed the world; turning almost any water, regardless of biological or toxicological contamination, into an affordable source of potable water; producing much of the world's energy needs from clean, renewable sources; developing energy-efficient, "green" manufacturing processes; remediating existing environmental damage; perhaps even enabling the blind to see, the deaf to hear and the lame to walk—no longer seem out of reach.

Anticipated Developments

It is both intuitive and reasonable to conclude that technologies that are powerful enough to deliver such extraordinary results also could be accompanied by potential negative developments as well, either accidentally—or through intentional creation of hostile applications.

Such power for good and for ill increases the importance and imperative of addressing such developments across institutions to ensure society maximizes the benefits of these new technologies while minimizing their downsides. Increasing globalization—trade, industrial, scientific, technological, capital, and workforce—magnifies the challenge, adding new layers of complexity given different value systems, states of economic development, geo-political alignments and interests, and economic models.

NBIC technologies are not simply about working at the margins of existing technologies, producing only incremental improvements, even though incremental improvements will emerge along the way. Convergence of these technologies has the potential to be disruptive, perhaps on a scale beyond what society has seen before. These technologies are likely to bring about and require organizational and broader social change by generating new science, new technologies, new industries, new manufacturing processes and capabilities, new services, new skills and knowledge, and destroying academic-disciplinary, industrial and governmental silos.

For individuals, NBIC technological convergence could result in the elimination of some jobs (as well as of companies and perhaps entire industries), the creation of new jobs and new occupations, and the need for additional education and "reskilling" on an ongoing basis. For companies and industries, NBIC convergence could result in the elimination of existing companies and industries and the emergence of new ones; cause substantial changes to and investments in private sector research and technology base, strategic partners, and even geographical location; and bring about changes in the occupational mix. For regions, NBIC convergence could result in the loss of existing industries and clusters and the opportunity to grow new clusters, with the attendant loss of revenues, social expenses (including unemployment compensation, infrastructure investment, retraining programs). For the United States, NBIC has implications for global economic, military, scientific, and technological leadership, national defense and homeland security, economic growth and jobs creation, and the American standard of living.

Developing technologies with such extraordinary implications requires a holistic approach that brings together all stakeholders and incorporates knowledge gleaned from fields, such as the life and physical sciences, engineering, business, economics, medicine, history, sociology, anthropology, ethics, theology, and political science.

Some around the world have called for a slow-down or outright moratorium on research in some of the NBIC technologies. Given the level and scope of investments, and the economic and societal potential, it is inevitable that research, development and commercialization of these technologies will occur. The question of who will lead the world in their development and commercialization and whether the research will be conducted in an informed and responsible manner so that society can maximize the benefits while minimizing the risk is still to be determined.

Research and Evaluation Methodologies

Given that the convergence of the NBIC technologies may profoundly change the way we live and work, there must be a strong, ongoing, open and honest dialogue among all stakeholders. Fundamental to this process will be scientific research to understand and quantify risks associated with nanotechnology materials, products, and processes. Ideally, those involved in the development of technologies would be mindful of and address societal and ethical implications as an essential part of the development process. At least in the beginning, models and analogies will be necessary to advance discovery and commercialization by applying new paradigms or lessons learned in one discipline to others.

Communication with the public on NBIC technologies is an important component of bringing such products and services to the marketplace. But the communication must flow in both directions, both to and from the public, in order to provide a vehicle for the expression of interests and concerns by those outside the scientific and engineering community.

"The public" is not a single homogenous group; there are multitudes of "publics" with a spectrum of views, shaped by a variety of factors, including age, sex, race, marital or parental status, geographic region, political affiliation, education, profession, religion, national origin, and economic ideology. Each public brings different knowledge, perspectives, expectations, hopes, desires, capabilities, interests, and fears to the table. Effective communication requires an understanding of the underlying foundations of the thinking, values, and belief systems that contribute to the perspectives of these publics. Knowledge of the various public perspectives will provide a foundation for more effective communication and will enable more informed participation on the part of these publics in the relevant discussions. Knowledge of public opinion also will improve the ability to identify and address the unique and common concerns of each group. It also will be necessary to foster understanding of science and engineering and the role they play in our economy, job creation, national defense, public health, and other areas of our lives. As there is no one "public," communications need to address the varying levels of understanding and concern of different stakeholder groups.

A unique challenge facing the NBIC converging technologies is the transference of negative qualities (real or perceived) associated with one of the technologies to the others and to their combinations in the perceptions of the public, which is essentially a case of guilt by association. For example, negative perceptions of genetically modified organisms (GMOs), human cloning, and stem cell research could be transferred from the biotechnology arena to the NBIC arena in the convergence of biotechnology and information technology, even though the new field and its products may be unrelated to the other work.

The multidisciplinary nature of the NBIC technologies requires increased communication between researchers in a variety of technical specialties. In addition, products that incorporate NBIC technologies may require communication across industry sectors and company divisions that traditionally did not interact. Fostering communication between and among these constituencies will be instrumental in advancing NBIC research and enabling its commercialization. The challenges to communication, however, can be significant.

NBIC research is "boundary work," requiring the construction and deconstruction of languages between communities. To foster effective communication among the scientific, technological, engineering, and business communities, individuals in the various groups must develop an ability to work across multidisciplinary "trading zones" to enable effective use of expertise and to improve the ability to interact with policymakers and the general public.

In many scientific and technical specialties, unique "languages" have emerged to describe and explain characteristics, processes, and phenomena that are unique to a field. While scientists and engineers speak a common technical language, significant variations can impede communication. A single term used by professionals of one discipline might mean something different to those from another discipline; conversely, different terms may be used to describe a common concept, phenomenon, or process. Convergence of science and technology requires convergence in the language employed to communicate them.

The power of NBIC technologies offers the potential for enormous societal benefits. The same power, combined with the uncertainties of working at the leading edge of technologies that demonstrate novel properties, creates significant questions of how to assess and manage risks. Thus there are needs for better models for risk analysis, characterization, and quantification; for more effective models to do risk-based cost-benefit analysis, especially in the public-policy sphere; and for improved dissemination of "lessons learned" in risk management.

NBIC technologies may contribute greatly to human health, longevity, and the easing of medical hardships, pain and suffering, and handicaps. Some NBIC applications, however, especially in the medical field, raise challenging legal and ethical issues. Researchers, for example, are creating new human-computer interactions and brain-machine interfaces that are generating new insights into brain function. These studies could impact the evaluation of an accused person's capacity to understand right from wrong, which is essential to the legal definitions of mental competence and criminal intent. Related technologies could help to restore lost brain functions or lead to treatments for mental illness or conditions such as epilepsy. Such research, however, raises important questions, such as, what kinds of treatments are appropriate, at what levels, for whom and under what circumstances?

Research, Education and Infrastructure Development

The revolutionary applications that may emerge from the NBIC technologies could have substantial implications for public and private institutions. Company and industry structures may need to change to bring together the knowledge, skills, and capabilities needed to bring NBIC technologies to market. Governmental organizations—policy, regulatory, legal—will need to adapt to changes brought about by NBIC technologies and to move at the pace of technological change. Academic institutions will need to overcome structural and process barriers that inhibit multidisciplinary approaches, research proposals, and conduct of work. All of these institutions will need an increased ability to evolve in an environment of continuous change, reducing human resistance to change, and breaking down the barriers between traditionally distinct realms of science and technology.

NBIC technologies may pose new risks associated with exposure to nanoparticles, environmental and ecological degradation from intended or accidental release of nanoparticles and bioengineered organisms (bacteria, viruses, plants, and animals), workplace exposure to nanoparticles, and compromise of personal information/ invasion of privacy. Nanotechnology-enabled ubiquitous sensors, computing, and information sharing are likely to raise a variety of legal privacy issues. Research is needed on synergistic interactions between nanomaterials and man-made and naturally-occurring compounds in the environment, transport of nanomaterials in the environment, transport from one medium to another, one organism to another, and transport from organism to the environment and vice versa. This knowledge is a foundation for regulatory decision-making covering development, manufacturing, use, disposal and reuse of NBIC products and materials.

The NBIC converging technologies are expected to pose challenges to the regulatory system. Regulatory agencies, in conformance with their individual missions, focus on specific areas, including food, drugs, environment, export controls, working conditions, banking, insurance, broadcasting, telecommunications, transportation safety, and nuclear power. The multidisciplinary nature of NBIC technologies blurs the boundaries between technical disciplines and industry sectors, and this blurring will have significant implications for regulatory bodies, such as the need for a substantial investment and shift in agency workforce education and training programs. NBIC technologies and products will require regulators to be cross-trained in areas outside of their core expertise, to expand their ability to work in multidisciplinary teams, and to engage in continuous skills upgrading to keep pace with rapid advances in technology. Although government agencies believe that their current regulatory authorities are adequate to cover nanotechnology-enabled applications and products that are emerging, future advances may necessitate legislative changes to agencies' mandates and spheres of authority and/or a restructuring of the regulatory review and approval process.

NBIC technologies have great potential to contribute to achieving regulatory missions, for example, protecting and enhancing human health (FDA); reducing pollution (manufacturing processes that produce fewer toxic wastes and by-products, more efficient manufacturing processes that reduce energy consumption and associated pollution, clean, renewable energy sources/storage technologies), and restoring the environment through remediation technologies (EPA), creating safe and healthy working environments (NIOSH), and ensuring safer consumer products (CPSC).

It is worth exploring proactive, non-regulatory approaches to achieving regulatory objectives. Industry might consider, for example, developing the means to share data on environmental and health effects so as to expeditiously identify and manage risks associated with these technologies. Representatives from industry, including manufacturers and insurers, also have expressed concern about uncertainties regarding regulation and safety that could lead to costly losses of R&D investments and the inhibition of future R&D investments. Expanded cooperation between industry and government would reduce such uncertainty thereby better enabling the achievement of regulatory objectives, while at the same time enhancing innovation and societal benefits.

NBIC technologies will bring new challenges to existing law, the educational preparation of legal professionals, and the legal infrastructure. NBIC technologies will challenge the ability of the courts and juries to adjudicate as long as risks remain undefined, thereby calling into question the current standards of proof. Business and legal decisions surrounding NBIC technologies may be complicated further by the lack of harmonized international laws, regulations, and standards.

NBIC may increase the importance of professional judgments of scientists and engineers in legal cases. In many legal cases today, both plaintiff and defendant bring credentialed scientific and engineering experts to court whose testimony may be contradictory to each other. Jurors often lack knowledge or even the underlying skills to learn about these complex technologies. NBIC technologies are likely to be quite complex and impenetrable to non-scientific jurists and jurors. Jurists, attorneys, and jurors will need education and instruction in the language, science, and technology of NBIC.

Action Recommendations

Technological convergence has the potential to achieve benefit for all human beings, if it does not become ensnared in unnecessary complexity, uncertainty, and public alienation. Therefore, actions, such as the following, should be undertaken:

- Research and education should be promoted about best practices in organizational design, so that organizations such as corporations and universities will have greater flexibility to change and be faster in making beneficial changes.

- To achieve a regulatory environment that protects the public while encouraging innovations, there should be greater coordination and consistency across Federal regulatory agencies, expanded coordination between Federal and state regulators, and efforts to establish common regulatory frameworks with other nations.

- Federal agencies should engage in research to understand, quantify, and mitigate risks to human health and well-being that arise from convergence of NBIC technologies.

- Government should experiment with new mechanisms for agencies to engage the public to provide a citizens' perspective.

- Global frameworks should be considered that offer common parameters for research, harmonization of regulations, and market access that could expedite the development and commercialization of beneficial NBIC technologies by reducing risk, creating transparency, and contributing to a level playing field for all competitors.

References

1. M. C. Roco, W. S. Bainbridge, *Converging Technologies for Improving Human Performance: Nanotechnology, Biotechnology, Information Technology and Cognitive Science*, Dordrecht: Springer (formerly Kluwer) (2003).

2. M. C. Roco and C. D. Montemagno, eds., The Coevolution of Human Potential and Converging Technologies, *Annals of the New York Academy of Sciences* **1013**, New York: New York Academy of Sciences (2004).

THEME 6: NATIONAL SECURITY AND SPACE EXPLORATION

Moderators: Delores Etter (U.S. Department of Defense) and James C. Murday (U.S. Naval Research Laboratory)

Contributors: Grant Black, George Borjas, Martin Carr, Minoo Dastoor, Linda Goldenberg, Kwan Kwok, Cliff Lau, John T. Neer, Ron Oaxaca, Judith Reppy, W. M. Tolles, Scott McNeil, Keith Ward

Introduction

Nanotechnology represents a number of scientific advances that provide new materials and advanced systems. These materials and systems will affect almost every phase of activity involved with advancing national security and space exploration [1, 2, 3, 4, 5]. Typically when NASA or DOD begins a new technology program, concern is centered on the performance benefits, cost of development, time for development and new opportunities to be enabled. Direct societal impact only becomes part of the process as the development or ultimate use of the technology directly affects the health or well-being of humans or other living creatures. Nanotechnology will enable NASA and DOD to build future systems with many advanced features. We need the capability for robotic systems to operate in dangerous environments without the high cost of continuous human control. However, as we develop new nanotechnology to accomplish these goals, we must also proactively establish policies and guidelines to assure the ramifications of these technologies and systems remain socially acceptable to the general public [6].

Current State of Knowledge

Maintaining security for citizens is one of the first and most critical requirements of government. The first step, gaining information through sensors, relates to the initial assessment of a situation, whether called "surveillance" or "being aware." Accurate and high-quality information contributes immeasurably to swiftly and effectively managing or mitigating potential or imminent risks and hazards. Assessing the information with decision aids to enrich and enlarge the scope of understanding is the second step, followed by the use of increasingly complex, multifunctional systems to act on the new knowledge. As we have seen in recent engagements in the Middle East, we also need advanced technologies to facilitate communication and understanding among a populace.

If it is necessary to employ force, the option to resort to weapons is considered. More accurate delivery of force, with less collateral damage, has been a trend over the last decade. The above trends in sensors and information technology, as well as the availability of stronger, lighter-weight structural materials and, to some extent, reliable explosives and propellants that release greater energy will enable this. Platform development is dependent on all of the above developments, especially on stronger, more durable materials for aircraft, armor for vehicles and ships, and submarines and satellites that are less vulnerable to corrosive environments. An important capability that emerges with miniaturization is that of unmanned vehicles, presently important for gathering information (in the sensor category again), but potentially also a means for controlled delivery of necessary munitions. As the human is further isolated from the application of force, appropriate safeguards should be in place to ensure that the United States meets its responsibilities under international treaties and laws of warfare.

Although regrettable, the application of force frequently involves injuries to personnel, and much can be done to minimize these occurrences and their impacts. Protecting the soldier is envisioned through improved sensors, shielding, information, and communication capabilities, monitors for health and body condition, and even clothing that is constructed to respond to injury automatically. The war-fighting environment involves adverse chemical and biological conditions that must be monitored and to which rapid responses are essential. These represent key opportunities for developments in nanotechnology. It is important to note that these same key opportunities have application to other domestic contexts that involve dangerous and hostile environments. Emergency personnel responding to natural disasters such as chemical spills, floods, hurricanes, and forest fires will clearly benefit from these technological advances.

It is likely that there will be contentious issues in the area of defense applications of nanotechnology. A policy of open discussion is needed to reassure public opinion, especially with respect to the safeguards as discussed above. Further, there is a considerable challenge posed for international security by the dual-use aspect of nanotechnology. Attention should be paid to international efforts that could minimize an arms race.

Anticipated Developments

Nanotechnology will clearly have wide-ranging impact on technologies in support of homeland security and space exploration. Especially important over the coming decade will be exploitation of the physical, chemical and biological properties of nanoscale building blocks. A good case can also be made for a near-term priority on developing directed hierarchical self-assembly of multifunctional systems. Over the longer term of 20 years, we can expect to see a diverse variety of commercially available, high quality, affordable nanoscale building blocks. Beginning now and extending out at least two decades, both national security and space exploration will benefit from large scale nanotechnology-enabled computing for astronomical, atmosphere, ocean, and earth system models.

A number of distinct research and development challenges can be identified as addressable via nanotechnology. Their solution would achieve important security and exploration goals. High-speed conversion of large data sets into meaningful information could be achieved through advances in quantum computing, parallel processing, and holography. Advanced robotics—high performance and energy-efficient—would make smart unmanned platforms possible for deep space exploration and combat vehicles with minimal human risk. Nanotechnology-enabled distributed sensing for real-time and continuous surveillance would be useful for defense against terrorism. Netted, broadband, secure communication would provide unimpeded ability to communicate, even during natural or man-made disasters. Composite materials with a high strength-to-weight ratio could facilitate very high-performance space launchers and fighter aircraft. Better control of failure mechanisms through the design of materials and system components at the nanoscale would reduce life cycle costs of both military and space equipment, for example, by reducing maintenance.

For the individual soldier on the battlefield, nanotechnology-enabled physiological sensors could constantly monitor vital signs and warn of exposure to chemical or biological warfare agents. An *active uniform* could adjust for environmental stresses, provide camouflage that matches changing background and lighting conditions, and even provide first-aid casualty response. Components manufactured from nanoscale-engineered materials could be lighter, allowing a solider to carry equipment with greater functionality. Perhaps it will be possible to achieve much greater energy density in portable power sources. The soldier will enjoy clear awareness of surrounding dangers and resources, with local information processing connected to netted communications, with essentially weightless electronics embedded into uniforms that require little electric power. Augmented-reality or virtual-reality learning that is tailored to the individual can provide effective training and preparation of the soldier before entering the battle area.

Over the coming decade, developments in four areas of nanotechnology are likely to be of value in space applications:

1. Development of composite matrices incorporating nanofibers or other nanoscale components may lead to programmable materials for multifunctional structures.

2. Advances in nanoelectronic components, nanoscale assembly, and systems architecture may lead to computers that achieve high capacity with little power consumption.

3. Nanoscale spacecraft components for harsh environments, combined with nanoscale sensors and instruments, could enable adaptive microspacecraft.

4. Nanotechnology systems for human health monitoring could be developed on the basis of biomolecular signatures of health conditions, molecular

imaging, signal amplification and processing, and possible self-assembly of mechanisms.

Fifteen to 20 years in the future, research and development in these areas could lead to adaptive airframes built from smart skin materials, highly intelligent nanoelectronic space probes, integrated smart nanotechnology sensor systems that tolerate radiation and high temperatures, and systems for diagnosing and treating human injuries and illnesses in flight.

Research and Evaluation Methodologies

The nanoscale characterization, modeling, and fabrication tools that have emerged have stimulated initial developments in nanotechnology, and a large number of technologies dependent on nanostructures are envisioned. In almost every phase of operations required for national security and for space exploration, nanotechnology provides advantages in the ability to gather, communicate, digest, and act upon information with advanced sensors, and to take requisite action with platforms that will have augmented capacity. From a societal perspective, there are important issues to address if the United States is to be in a position to fully exploit the potential opportunities.

An expected decline in the number of foreign students, combined with a rise in demand for nanotechnology-related scientists and engineers, implies that the United States must produce an increasing number of domestic scientists and engineers. Concerns about the declining number of U.S. citizens choosing science education and careers must be addressed. Evidence is needed on what has influenced these students to opt out of science. Factors such as reduced labor market opportunities, the prevalence of foreigners willing to receive lower wages and the lack of scientific awareness may have contributed to this trend. Research must address questions such as: What is the societal impact of having foreign nationals participate in nanotechnology-related education, labor markets and research? What is the long-term impact of having fewer foreign nationals contributing to U.S. science, particularly in nanotechnology fields? Should international cooperation in nanotechnology be encouraged to reduce the risk of global misuse of nanotechnology and increase access of U.S. engineers to leading nanotechnology R&D?

Looking 10 years ahead, the U.S. university system must determine appropriate levels of training and the most efficient methods of delivery for nanotechnology-related training to provide an adequately trained scientific workforce. More accurate information on the expanded size of the nanotechnology workforce, its required skills, and the scope of nanotechnology applications is needed to determine the most efficient educational outcomes. For example: What are the roles of associate, bachelor, masters, and doctorate training, as well as short-term interdisciplinary nanotechnology-focused training programs? What are the most efficient methods of implementing nanotechnology in the educational system?

A potentially productive focus for the education question would be to examine the massive, but largely untapped, reservoir of potential and skills acquired by young people through interaction with games and entertainment. Physical and cognitive skills like hand-eye coordination and reaction acquired through computer games could be redirected into different contexts such as combat and learning. For example, controls for weapons and munitions could be designed to be similar to controls for games, thereby matching existing human skills and capabilities acquired from gaming. This would minimize training and retraining, and maximize the existing human potential pool. Similarly, educational technology and curricula could incorporate games and nanotechnology content, having the effect of integrating gaming skills acquired in an entertainment setting with the educational context and science content. The point is that it may be useful to step outside the traditional educational "box" separating education, evaluation, and curriculum from other dimensions of life, and shift focus to gaming and entertainment as a potentially useful site for educational development. This also relates to "cognitive readiness," as young people may be more cognitively ready than recognized for new technologies.

Research, Education and Infrastructure Development

The national priority on nanotechnology has significant implications for U.S. science and engineering education, particularly given national security concerns related to the global adoption of nanotechnology and to the potential consequences for U.S. competitiveness. Two specific educational issues are:

1. the development of an adequate U.S. nanotechnology workforce in the short, medium, and long terms

2. public awareness of the importance of nanoscale science and technology

The adoption of nanotechnology over the coming decades will initiate an increase in demand for scientists and engineers both globally and in the United States. Approximately 25 percent of current U.S. Ph.D. degrees are awarded in nanotechnology-related fields, and this will likely continue to increase. To meet this growing demand, and to enable U.S. competitiveness in nanotechnology, we will need a sufficient supply of scientists and engineers who have the necessary skills to competently develop and use nanotechnologies. The pipeline includes both domestic and foreign scientists and engineers. Concern exists, however, regarding the sufficiency of domestic supply. The proportion of foreign students in the sciences, including nanotechnology-related fields, has increased dramatically in the past 20 years. In 1980, approximately 18 percent of Ph.D. recipients in the United States were non-U.S. citizens, compared to nearly 33 percent in the 1990s.

Many nations are beginning to invest in research and development of nanotechnology, so employment opportunities outside the United States may begin to lure U.S.-trained foreign students away. Recent tightening of immigration policies will continue to reduce the supply of foreign students to the United States. This will occur in two ways:

1. by reducing the likelihood that foreign-born graduates of U.S. institutions stay here

2. by discouraging foreign students from pursuing a U.S. education

Evidence suggests that foreign-born scientists and engineers have made disproportionate, positive contributions to U.S. science relative to natives [7]. Therefore U.S. nanotechnology-related science may be negatively impacted by the likely reduced contributions of foreigners in the U.S. scientific enterprise. However, the risk of potential negative effects due to the education and training of foreign nationals in the United States could be simultaneously reduced.

Beyond the security implications of foreign students, national security should consider the ease with which information is transferred, particularly in the academic environment. At the present time, knowledge derived from basic research is distributed relatively freely, with some controls (export/import) coming into play for results from applied research. In the past 20 years, transfer of information has become less costly and more frequent, due in large part to technological inventions such as the Internet and email. Collaborative networks in the sciences have expanded in size and grown increasingly international [8]. Research is needed on the magnitude of risk this relatively free exchange of ideas has on U.S. competitiveness and security. There are advantages and disadvantages to the United States in the exchange, so care must be taken to develop a balanced perspective. Otherwise, the U.S. Government could inadvertently act to its disadvantage.

It is appropriate to create a communication strategy to keep the public informed of a cross-section of research and development activities. There are substantial risks in not doing so, such as the risk of negative reaction to speculation, partial information and misinformation at the national and international levels. But more importantly, public awareness of interesting and significant research and development will stimulate the best minds to see other potential applications, and motivate these minds to pursue science education and careers consistent with their visions. Viewing the populace as more than citizens with rights, but also as people with capacity and responsibility, opens new vistas.

In order to maximize this capacity, citizens must be informed and engaged. This calls for a communication strategy that motivates citizens and enables them not just to understand nanotechnology developments in national security and space exploration, but also to envision potential applications in their personal worlds and local contexts. For example, sensors, materials, decision aids, and computer and communication capabilities developed within a military context will benefit police, firefighters, paramedics, and other emergency personnel responding to chemical spills, floods, hurricanes, blizzards, forest fires, and search and rescue calls. Technological advances will contribute not only to response, but also to preparedness and mitigation, and to communication and coordination during domestic crises.

To a great extent, technology development in the context of national security and space exploration involves enabling technologies with broad application. Sensor technologies have potential application in environmental and healthcare monitoring, materials in recreational activities, and computer capabilities in business and industry. Public awareness and an informed citizenry are essential elements in the effort to expand applications, and in maximizing the capacity and visions of responsible citizens.

Action Recommendations

Programs must be established to provide awareness of the importance of nanotechnology for societal benefit and to stimulate interest among students in pursuing science education. A strong role exists for national agencies, such as the NSF, to solicit, fund and disseminate such programs at a national level.

- Federal support should be provided for K-12 curriculum development and educational programs on nanotechnology awareness, understanding, and importance to stimulate interest in science and attract dedicated students.

- Economists and social science researchers should develop an accurate estimate of scientific workforce needs (market size and types of skills) and a timeline of nanotechnology adoption and needs, based on input from industry and universities.

- To meet current and short-term labor needs, the government should support the implementation of retraining programs to equip underutilized scientists and engineers, in areas with poorer labor market prospects, with nanotechnology-related skills.

- Leaders in science, education, and government should develop a communication strategy to keep the public informed of representative and fundamental developments in nanoscience and nanotechnology.

It is also recommended that substantial investment be made to explore yet unanswered research questions related to the implications of nanotechnology on national security, including:

- Should international cooperation in nanotechnology be encouraged to reduce the risk of global misuse of nanotechnology and increase U.S. access to leading nanotechnology R&D?

- What is the societal impact of foreign nationals in nanotechnology-related education, labor markets, and research?

- What is the long-term impact of fewer foreign nationals contributing to U.S. science, particularly in nanotechnology fields?

- What are the most appropriate levels of nanotechnology training needed in the short and long terms?

- What are the most efficient methods of implementing nanotechnology in the educational system?

References

1. D. Ratner, M. A. Ratner, *Nanotechnology and Homeland Security*, Upper Saddle River, NJ: Prentice Hall (2004).

2. M. A. Ratner, D. Ratner, *Nanotechnology: A Gentle Introduction to the Next Big Idea*, Upper Saddle River, NJ: Prentice Hall PTR (2002).

3. [DOD researchers provide] A look inside nanotechnology, *AMPTIAC Quarterly* **6**, 1-68 (2002).

4. M. C. Roco, W. S. Bainbridge, eds., Section E in *Converging Technologies for Improving Human Performance: Nanotechnology, Biotechnology, Information Technology and Cognitive Science*, Dordrecht: Springer (formerly Kluwer) (2003).

5. Miziolek, et al., Symposium on Defense Applications of Nanomaterials. Paper read at 221st American Chemical Society National Meeting, Division of Industrial and Engineering Chemistry, 1-5 April, San Diego, CA (2001).

6. M. C. Roco, W. S. Bainbridge, eds., *Societal Implications of Nanoscience and Nanotechnology*: National Science Foundation Report, Arlington, VA: National Science Foundation (2000), also Dordrecht: Springer (formerly Kluwer) (2001) (available at http://www.wtec.org/loyola/nano/NSET.Societal.Implications).

7. P. Stephan, S. Levin, Exceptional contributions to U.S. science by the foreign-born and foreign-educated, *Population Research and Policy Review* **70**, 59-79 (2001).

8. J. Adams, R. Clemmons, G. Black, P. Stephan, Patterns of research collaboration in U.S. universities, 1981-99, submitted to *The Economics of Innovation and New Technology*.

General References

J. Baker, R. Colton, H. S. Gibson, M. Grunze, S. Lee, K. Klabunde, C. Martin, J. Murday, T. Thundat, B. Tatarchuk, K. Ward, *Nanotechnology Innovation for Chemical, Biological, Radiological, and Explosive (CBRE): Detection and Protection*, Workshop Report (available at http://www.wtec.org/nanoreports/cbre/).

Chemical Industry Vision 2020 Technology Partnership, *Chemical Industry R&D Roadmap for Nanomaterials by Design: From Fundamentals to Function*, Workshop Report (available at http://www.chemicalvision2020.org/nanomaterialsroadmap.html).

G. Black, P. Stephan, Importance of foreign Ph.D. recipients to U.S. science, in R. Ehrenberg and P. Stephan, eds., *Science and the University*, Madison, WI: University of Wisconsin Press (to be published).

G. Borjas, *An Evaluation of the Foreign Student Program*, Washington, DC: Center for Immigration Studies (2002).

G. Borjas, *Heaven's Door: Immigration Policy and the American Economy*, Princeton, NJ: Princeton University Press (1999).

P. Stephan, G. Black, J. Adams, S. Levin, Survey of foreign recipients of U.S. Ph.D.s, *Science* **295**(22), 2211-2212 (2002).

M. Finn, *Stay Rates of Foreign Doctorate Recipients from U.S. Universities, 1997*, Oak Ridge, TN: Oak Ridge Institute for Science and Education (2000).

THEME 7: ETHICS, GOVERNANCE, RISK, AND UNCERTAINTY

Moderators: Vivian Weil (Illinois Institute of Technology) and Rachelle Hollander (National Science Foundation)

Contributors: Carol Lynn Alpert, Arthur Caplan, Daniel Goroff, Sheila Jasanoff, Daniel Jones, Frank Laird, Bruce Lewenstein, Jane Macoubrie, Robert McGinn, Julia Moore, Deb Newberry, Philip Sayre, Albert Teich, John T. Trumpbour

Introduction

Policymakers and scholars must achieve an engaged understanding of issues of ethical and social responsibility, with regard to individuals and institutions and developments in emerging science and technology. Progress will require genuine respect for interdisciplinary discussions about the ethical and social dimensions of nanoscale science, engineering, and technology. On the way to that goal, a better understanding of systems complexity, and uncertainty is required so as to address the unusual diversity, complexities, and uncertainties of the nanotechnology area. Also important is a better understanding of how research directions get set and revised, including an understanding of the roles played by various government agencies and government interagency initiatives. Research should be directed at gaining knowledge in each of these areas.

Current State of Knowledge

Nanotechnology area is extremely diverse and complex. The importance of this area for producing enabling technologies and tools is clear. Little is currently known, however, about the broader social context of the emerging technology. Research, for instance, is needed to better understand the roles of institutions that will play a role in nanotechnology development and the roles of those that may be involved in control or regulation of nanotechnology developments.

Some efforts to advance understanding of the nanotechnology area across disciplines and to engage with members of the public are becoming visible, including efforts in the social sciences and humanities, as well as in the engineering and physical science disciplines, e.g., the University of South Carolina programs. Although these beginning efforts are not yet advanced enough for assessment, they do provide suggestive models and pilots.

The workshop participants in this breakout group reported that research generally shows that the "news model" of public involvement, in which technical experts and the media impart information to a passive audience, fails to bring about an informed public. Information systems that allow two-way conversation are much more appropriate for new technologies. The context of such conversation is relevant to a public's uptake of knowledge, and lay knowledge is not to be ignored (the public is not made up of empty vessels). Public participation and democratic processes, which allow the public a measure of control, are integral to meaningful conversation. Issues of power and trust are at the core of public debates and controversy.

While hyperbole in promotion is unavoidable and probably necessary, it inevitably evokes doubts. Examinations of this phenomenon are needed.

Because innovative technologies bring about unintended consequences, it also will be more fruitful to try to shape the future by building institutions that can adapt to emerging issues while preserving core values, rather than try to predict the future.

Early findings indicate the toxicity of certain nanotechnology products, with noteworthy damage to the lungs of rats and mice under certain conditions from a form of carbon nanotubes [1]. The chief conclusion from the studies reported to date is that more extensive follow-up studies are needed [2]. Social scientific studies of risk indicate that lay perceptions of risk differ from the perceptions of experts. Scientists' perceptions of risk often overlook social risk. A remaining question is whether there is inevitable tension between lay and expert perceptions. Can risk studies elucidate these differences and tensions in ways that will be useful for understanding response to new developments in nanotechnology?

In fiscal year 2000, the National Science Foundation began tracking projects from the Nanoscale Science and Engineering (NSE) program that contribute to the goal of understanding the societal and ethical implications of nanotechnology. Further information on these projects is available at the NSE website: http://www.nsf.gov/crssprgm/nano/.

Of special note are two Nanotechnology Undergraduate Education (NUE) projects and two Nanoscale Interdisciplinary Research Teams (NIRT) projects that were funded in fiscal year 2003. The NUE projects at Rochester Institute of Technology (0304308, Principal Investigator Paul Peterson) and at Michigan Technological University (0304439, Principal Investigator John Jaszczak) aim to integrate studies of the social and ethical dimensions of nanoscience and nanotechnology into undergraduate curricula and to provide models for other colleges and universities. The two NIRT projects have quite different focuses. "From Laboratory to Society: Developing an Informed Approach to Nanoscale Science and Technology," a program at the University of South Carolina that is under the direction of Davis Baird (0304448), examines concepts of understanding and control as they influence scientific and technological developments as well as public involvement and reactions to this emerging field. The "Science and Commercialization Nanobank, Database and Analysis," a program at the University of California Los Angeles, led by Lynne Zucker (0304727), is building an integrated database that will be made available as a public, web-deployed digital library called NanoBank.org. It will be useful to researchers, firms and investors, policy makers, and scientists and engineers working in the field. They will be able to use its matching and searching capabilities to understand developments and evolution in the networks and activities that make up this emerging field.

Anticipated Developments

Specific developments anticipated in 10 years include the creation of one or more online bibliographies, which will comprise accessible and usable website resources for readings and information on ethics and nanotechnology. These will be integrated into continuing education for entrepreneurs, scientists, and engineers studying the ethical and societal implications of nanotechnology. More broadly, there will be multidisciplinary endeavors to educate physical scientists, engineers, social scientists, and humanities scholars on nanotechnology and society, with interdisciplinarity being made more central to science and engineering education. One beneficial outcome will be the training of a cadre of junior investigators and postdocs, working on the ethical and societal implications of nanotechnology.

One can also foresee the development of centers conducting ongoing critical, reflective research on science, technology, and society, including comparative studies; research in the nanotechnology area would fall within this body of work. Research will include investigations about the definitions of humankind ethically, philosophically, and religiously, all in relationship to science and technology.

Research and Evaluation Methodologies

One prerequisite for research on ethics, societal implications, and nanotechnology is the creation of structures in which scientists, ethicists, humanities scholars, and social scientists work together. Another prerequisite is the organized skepticism that is characteristic of science. Methodologies will be developed and refined as part of the coevolution of the physical, biological and social sciences, and humanities in the nanotechnology area.

One method is to test practical efforts to engage the public with openness to many models and contexts. Another is to conduct comparative studies, examining science and technology in a global context and from a range of cultural perspectives. Also valuable would be historical studies of earlier innovative technologies, such as biotechnology, and prior government interagency initiatives, such as the strategic computing initiative.

Fundamental knowledge about the origins and functions of interest groups is needed, and the requisite elements needed by each group to participate responsibly in technological development should be identified. One fruitful approach is to examine nodes of controversy, among informed and uninformed parties, with attention to cross-cultural differences. Another is to examine the analogies that illuminate the nanotechnology area, e.g., "the next industrial revolution," with attention to the use and impact of hyperbole. Related research topics are the concept of the public (a historical, voting public or a public engaged on the issue of nanotechnology), public or civic knowledge, social responsibility of individuals, organizations, and institutions, dialogic processes for discussion, forms of knowledge production and diffusion, and modalities of education and of governance (rhetoric). How does public knowledge develop and evolve? Scientists, philosophers, and policymakers should rethink this terrain.

Valuable contributions can be made by studying institutions with respect to their responsibilities for health and safety. Institutional maps can be used to help discern where responsibilities lie and to better understand regulatory policy. This approach would include comparative research on international styles of governance of science and technology, as well as public participation and decision making. Practical insights could be gained by research on specific efforts, such as the NIH Centers for Excellence in Ethics, with respect to training, research, and outreach, and collaborations with outside centers. A crucial research focus should be on national initiatives as vehicles for building support for scientific research. How does hype work, especially in the marketing of science? Who are the publics? What are appropriate deliberative processes (hype included)? What forms of knowledge production and diffusion are pertinent? What modalities of governance other than government regulation are relevant?

There is a range of relevant research in fields such as history of technology, bioethics, public understanding of science, communications, and science and technology studies. These fields offer agendas and theories that need consideration and augmentation. Additionally, interactive approaches are needed to engage citizens and young people in discussion of these issues and choices that face them. Governing institutions need to develop ways of learning and adaptation. Rather than focusing on predicting the future, we should recognize that individuals and institutions create it. The values we embed in ourselves and our institutions will influence the future.

Research, Education, and Infrastructure Development

A paramount educational initiative is the training of a cadre of postdoctoral fellows and graduate students who combine specialties in the social sciences and humanities with a knowledge of nanoscience and engineering. Research needs include the development of methodologies for determining the set of relevant societal impact issues and for determining the content of the information that the public needs. We also need theory about publics and communication with them, about how publics evaluate information, about the nature of publics' logics and knowledge, about how publics govern, and about responsibility between states and citizens compared with responsibility between organizations and citizens. We need tools for a more adequate understanding of governance by multinational corporations and private sector organizations that extend beyond the traditional focus of political scientists on governments. We need further examinations of trust and power and their interrelations, and an understanding of systems, complexity, and uncertainty, taking off from the literature on systems in the field of Science and Technology Studies (STS) and addressing the unusual diversity, complexities, and uncertainties of the nanotechnology area. Progress in these areas requires archival and communications infrastructures and access to data and databases.

Action Recommendations

Government, industry, and academia should create opportunities for conversation between nanotechnology specialists and members of the public, to forge shared standards of reasonableness. Genuine feedback from relevant publics to nanotechnology specialists might influence their decision to emphasize development in one area or to choose not to develop another area. This would empower the public, which is a requisite for moving forward responsibly. Also to be addressed are the characteristics of the culture of innovation, the propagation of enthusiasm, the function and ethics of hyperbole, and the creation of new institutional values, as well as the need for fashioning governing institutions that can learn and adapt in creating the future. Additional recommendations include:

- The need for institutional reforms should be evaluated, concerning university-corporate ties, transparency, secrecy and disclosure, privacy, suppression of information, contracts, and mechanisms of accountability (for example, the GAO) as they affect innovation and public trust.

- Projects should incorporate ongoing engagement of publics in deliberation and discussion about nanotechnology, developing infrastructures for balanced and inclusive public participation with many different, innovative models used to assure two-way interchange between nanotechnology engineers and scientists and their publics.

- Educational initiatives should aim to enhance critical thinking and provide structure and support for graduate and postdoctoral students; cross-disciplinary training and experiences; models for collaboration of physical, biological and social scientists, and humanists across disciplines; and for integration of social science and technical research.

References

1. D. B. Warheit, B. R. Laurence, K. L. Reed, D. H. Roach, G. A. Reynolds, T. R. Webb, Comparative pulmonary toxicity assessment of single wall carbon nanotubes in rats, *Toxicol Sci* **77**, 117-125 (2003).

2. P. H. M. Hoet, I. Brüske-Hohlfeld, and O. V. Salata, Nanoparticles-known and unknown health risks, *J. Nanobiotechnology* **2**, (2004).

3. J. Corn, ed., *Imagining Tomorrow: History, Technology and the American Future*, Cambridge, MA: MIT Press (1986).

4. J. Gregory, S. Miller, *Science in Public: Communication, Culture, and Credibility*, New York: Plenum (1998).

5. T. P. Hughes, *Rescuing Prometheus*, New York: Pantheon Books (1998).

6. J. Krige, D. Pestre, *Companion to Science in the Twentieth Century*, London: Routledge (2003).

7. M. C. Nisbet, B. V. Lewenstein, Biotechnology and the American media: The policy process and the elite press, 1970 to 1999, *Science Communication* **23**, 359-391 (2002).

8. V. Weil, Zeroing in on ethical issues in nanotechnology, *Proc. of the IEEE* **91**, 1976-1979 (2003).

THEME 8: PUBLIC POLICY, LEGAL, AND INTERNATIONAL ASPECTS

Moderators: Evelyn L. Hu (University of California, Santa Barbara) and T. James Rudd (National Science Foundation)

Contributors: Nila Bhakuni, William R. Boulton, Mike Davey, Michael Heller, Jim Von Ehr II, Sheila Jasanoff, Thomas Kalil, Robert McGinn, Sonia E. Miller, Julia A. Moore, Kesh Narayanan, Judith Reppy, Glenn Rhoades, Nora Savage, E. Jennings Taylor, George Thompson, W. M. Tolles, Raymond K. Tsui, Vivian Weil

Introduction

Nanoscience and nanotechnology are well recognized as crosscutting endeavors. The complexities of such broadly integrative fields are mirrored in public policy and legal issues. The breadth of the scope that nanotechnology encompasses, the multitude of the transformations possible, the relative youth of nanotechnology as a field, and the accelerating progress of its growth make public policy and legal issues critical and complex. These issues will profoundly affect the future development of nanotechnology and the nature of its impact within society. The focused attention and investment in nanoscience and nanotechnology on a worldwide level, and the global aspects of nanotechnology's transformative capabilities, point to the need for a careful examination of its development within an international context.

Current State of Knowledge

A large number of important public policy issues are raised by the National Nanotechnology Initiative, but three are clearly of great significance:

1. the overall funding level for the NNI

2. implications of nanotechnology for the developing world

3. environmental and human health effects of nanomaterials

Although there has been a significant increase in funding for nanoscale science and engineering, we are arguably still under-investing. At the same time, some of our competitors are investing much more in nanoscale science and engineering as a percentage of their GDP. Given the likely considerable "return on investment" associated with nanoscale science and engineering, the increased funding may be critical. For example, a large fraction of the increase in U.S. productivity in recent years can be attributed to dramatic reductions in the cost of storing, processing and transmitting information, coupled with changes in organizational and work practices that take advantage of new technologies. Nanoscale science and engineering may allow us to continue (or even accelerate) these trends for several additional decades, as today's technologies (e.g., silicon CMOS) begin to reach fundamental limits.

Agencies are still able to fund only a fraction of the meritorious proposals that they receive—sometimes 10 percent or less. Grant size and duration are often inadequate. Some grants are barely enough to support a single graduate student.

Researchers have identified a host of fundamental research questions and promising applications that we may not be able to cover adequately under the existing budget. For example, despite the broad range of environmental applications, including remediation, filtration, monitoring, and pollution prevention, the EPA has less than 1 percent of the total budget of the NNI.*

The potential value of nanotechnology for the developing world cannot be overestimated. Four billion people on the planet earn less than $2,000 per year. Every day, more than 30,000 children die of preventable diseases. At least 1.2 billion people lack access to safe drinking water. For the 2 billion rural poor, biomass (wood, crop residue, and dung) is still the dominant source of fuel. The indoor smoke from solid fuel is one of the top 10 risk factors for the global burden of disease, accounting for 1.6 million premature deaths each year. Nanotechnology research has the potential to play a considerable role in mitigating these problems. For example, nanotechnology could be used to create a low-cost "lab on a chip" for infectious diseases that are prevalent in developing countries, affordable, carbon-free sources of energy that are accessible to the rural poor, water filtration systems that increase access to safe drinking water, or inexpensive, accurate, real-time sensors that can help protect water and air quality.

Every beneficial technology has undesirable side effects, so close attention should be paid to impacts, including the environmental and human health effects of nanomaterials. As nanotechnology applications reach the commercial marketplace, the public should be confident that the government is taking appropriate steps to safeguard the environment and human health, while also allowing new technologies and new industries to flourish. The existing regulatory structure for the definition, classification, and hazard rating of new chemicals falls within the purview of the Toxic Substances Control Act (TSCA) at the EPA, which gives the agency the ability to track currently produced, imported, and newly manufactured chemicals [1]. The current method for the classification of compounds is by chemical formula and structure, thereby declaring macroscale and nanoscale compounds that have the same chemical structure as the same compound. However, these nanoscale materials can have different chemical, physical, electrical, electronic, and optical properties. Consequently, this strategy is currently under review within the Environmental Protection Agency Office of Pollution Prevention and Toxics, which has oversight of TSCA and related matters.

---

* Editors' Note: Other agencies funding environmental, health, and safety research include NSF, DOD, DOE, NIH, USDA, NIOSH, and DOJ. Total funding in this area for FY 2006 is estimated at $38.5 mullion.

Anticipated Developments

Laws and public policies will change, and they will shape the ways in which nanotechnology develops and how it changes our world. Changes could be global in nature, but they also could be felt locally, so that decisions and investments will need to be made at several levels and in many distinct domains, including internationally.

Commercialization of nanotechnology will be industry specific. For example, the semiconductor and electronics industry is intensely competitive with very small margins, and multinational Asian companies dominate many markets in this area. Additionally, debate continues over exclusive licenses by universities, especially with respect to the commercialization of drugs. The U.S. Government will need to contemplate the economic impact of the Bayh-Dole Act and consider the possible enactment of other legislation that can help to make American companies more competitive.

As nanotechnology and nanoscience converge with other sciences and technologies, and as advanced computing and human-machine integration speed forward, the resulting issues will cut across several legal practice areas. Some of these may include

- torts, due to the potential for personal injury from product misuse or mishap, whether intentional or negligent, and the trespass of nanoparticles

- environmental law, due to unknown risks of radically new technologies, such as the effects of inhaled manufactured nanoparticles, the release of buckyballs into the air and water and their effect on the environment and the food chain, and the likelihood of exposure to hazardous materials and potential subsequent toxicity to people and other organisms

- employment and labor law, with the potential for discrimination resulting from issues of equity, distribution, and access

- health and family law, as a result of genetic intervention capabilities, creation of artificial life forms, and stem cell research

- criminal law, through advanced DNA forensics

- constitutional law, protection of individual rights and equal protection as privacy rights, security and surveillance become more invisible through advances in computing, biometrics, e-commerce, sensing equipment, and Federal legislation

- international trade laws, trade regulation, customs, immigration, and cross-border jurisdiction all impact interstate commerce

- antitrust concerns, due to increased collaboration among competing parties

The societal implications of nanotechnology are not issues for the United States alone. According to Renzo Tomellini, head of the European Commission's Nanotechnologies and Nanosciences Unit, "Our scope is to help people, to serve people, to improve the quality of life for people, to improve industrial competitiveness, to protect or improve the environment, to support European policies.... Nanotechnology is a tool, an approach.... The interesting thing is that nanotechnology seems to be a very powerful approach to achieving these goals" [2].

The global flow of research and development in nanotechnology will be influenced by international arms control agreements and by corresponding threats of terrorism. The Biological and Toxin Weapons Convention (BTWC) and the Chemical Weapons Convention (CWC) outlaw the development of new agents and new means of delivery for biological or chemical weapons. The creation of new nanotechnology materials and components might well have toxic or dangerous effects, which would violate these conventions. Thus there is a need to monitor nanotechnology developments for treaty compliance.

In contrast to the international trade regime is the arena of national security where international agreements have been used to constrain the development and transfer of new weapons technologies that have been judged destabilizing or contrary to international law. Proliferation of weapons technology to developing countries, particularly those labeled "countries of concern," has been discouraged simply through mercantile means, such as through suppliers' cartels and through the outright prohibitions contained in various arms control treaties. These bans will apply to nanotechnology when it appears in the guise of improvements to weapons covered by arms control regimes. Such controls will apply to nanotechnology when it appears in the guise of improvements to weapons covered by existing arms control regimes. Like other technological advances, nanotechnology could be used in nefarious ways that are outside the scope of existing controls. Therefore, those working in the area of national security policy should be alert to any potentially harmful or negative use of nanotechnology.

Research and Evaluation Methodologies

Research in nanotechnology is now considered a strategic initiative in countries around the world and the outcomes of supported research will impact future developments in the global economy [3]. For this kind of international endeavor, which has widespread consequences for the participating country, a high level of coordination is recommended from the outset. As an historical example, in biotechnology in the 1980s and 1990s, the Organization for Economic Co-operation and Development (OECD), with its 30 member countries and active relationships with some 70 other countries, played a major role in bringing together scientific, technical, and national government experts to develop a common set of mutually agreed-on definitions, to foster information exchange among science and engineering professionals, governments, and industry (including regulatory and legal experts), to collect reliable basic data and statistics, to promote policies that encourage rapid understanding and appropriate diffusion of beneficial applications to society, and to discuss practical approaches for evaluating, harmonizing, and establishing national and international safety guidelines.

The United States should continue to support bilateral or regional programs, such as those undertaken between the NSF and the European Commission. Future topics for such programs could include the implications and applications of nanotechnology for the environment, how nanotechnology could address the sustainable energy challenge, and nanotechnology's potential societal (including legal) impacts. Survey research should be carried out to collect comparative data on public attitudes toward nanotechnology in key countries. Perspectives about priority applications, privacy and ethical concerns, and preferences regarding legal regimes and proposed safeguards would be extremely useful.

Multidisciplinary research would be facilitated by the development of global uniform definitions of terms and communication protocols. Social scientists, legal scholars, and policy makers must cooperate to determine the positive and negative effects of government regulation versus self-regulation and to identify the proper forum through which to address international implications.

Evaluation requires competent evaluators, so it will be necessary to invest in education and development in Federal agencies, regulatory bodies, and judicial systems. Similarly, it will be necessary to educate the scientist, engineer, and technologist to the science of business, ethics, and jurisprudence, and to educate the consumer to the facts of nanotechnology.

Research, Education, and Infrastructure Development

With science and technology's relentless advancements acting as centrifugal forces on society, today's civil justice system must be prepared to lead and help shape the new values, standards, and possible rules brought about by this enabling tool called nanotechnology. Through the integration of nanotechnology research and development, together with commercialization of applications, our current manufacturing processes, educational systems, business models, economic structures, and healthcare, environmental, defense, space, energy, and societal frameworks will be transformed. Current laws, regulations, policies, and judicial infrastructures will be directly impacted and challenged. While the need to address the legal and regulatory implications of nanotechnology is oftentimes mentioned in reports, workshops, and conferences, the broad complexity of the issues within the multiple clusters of legal practice areas cannot be overlooked. Nanotechnology impacts not only intellectual property, the commercialization and technology transfer of research results and products from laboratory to market, but also a wide integration of multiple legal practice areas.

Intellectual property rights are among the most significant legal aspects of nanotechnology. As noted in the U.S. Constitution, the purpose of the U.S. patent system is "to promote the progress of science and useful arts" [4]. The only President to receive a patent, Abraham Lincoln, said that the patent clause in the Constitution "...added the fuel of interest to the fire of genius, in the discovery and production of new and useful things" [5, p. 363].

More recently, the patent system has been described as the "[emphasis added] *primary* policy tool to encourage the development of new technologies" [6, p. 101]. The fact that the U.S. patent system plays a dramatic role in technological innovation can hardly be disputed. Studies have clearly indicated an explosion of the number of nanotechnology-related patents in recent years [8, 9]. In addition to the volume of patents and patent applications, indications of the growing complexity of patents issued today versus those of 20 years ago have been noted [7]. A key issue regarding nanotechnology will be the impact of the increasing number and complexity of patents on the patent system (i.e., statutory law, examination, and common law) and the ability of the patent system to effectively provide the incentive for technological innovation.

The term "technology transfer" has a multitude of definitions, but in this particular panel report it refers to the transfer of research results from universities and government research entities to the commercial marketplace. As U.S. Government agencies invest money in university research efforts for nanoscience and nanotechnology, they, along with companies and universities, are correct to contemplate the many ways that this research can be transferred into the commercial sector. Industry-university collaborations, patenting, the creation of start-ups, licensing, and industry-sponsored research are all different forms of "productizing" university research for the public benefit.

Patent applications are filed to provide industry with an incentive to invest in the development of the patented invention. As with any invention, universities should file patent applications on nanotechnology inventions when a temporary monopoly is necessary for the development of the idea. Many times in nascent technology fields such as nanotechnology, university discoveries will yield "platform technologies," which can be the basis of a start-up company. Nanotechnology will most likely be commercialized through highly technical established companies as well as through start-ups. The Small Business Innovation Research (SBIR) and Small Business Technology Transfer (STTR) programs are Federally funded mechanisms to support (existing or start-up) small businesses to bring nanotechnology into the marketplace.

Intellectual property (IP) becomes an important strategic component to a company's competitive advantage. In our litigious society, infringement suits can threaten a company's viability because the costs may be tremendous. As the IP landscape becomes more crowded and the possible commercial rewards expand, the legal strategy and competence of a company may determine its success or even its survival. Universities will have to consider the IP landscape as well, since they may be vulnerable to infringement suits in the pursuit of academic research.

Because nanotechnology encompasses many interdisciplinary applications, it is appropriate that the university community understands all potential applications of nanotechnology discoveries, and conducts a "fields of use," or sector-appropriate, analysis. Universities should make case-by-case decisions on whether to patent and in what countries to patent. Licensing strategy is important. Issues to be considered include whether to license exclusively, nonexclusively, or by field of use, and what diligence terms to include to ensure that the technology will be commercialized. The Bayh-Dole Act has strict regulations that give preference to small companies and substantial manufacturers in the United States for exclusive licenses while the government retains a non-exclusive license for government purposes. In this regard, SBIR and STTR companies are positioned to take advantage of IP emerging from Federal funding. As nanotechnology spans many interdisciplinary fields, commercialization efforts will require interdisciplinary teams.

A critical issue that affects the scope and vitality of nanoscience research is the infrastructure of talented and well-trained personnel that will carry out that research. The Association of International Educators (NAFSA) reported in January 2003 that increased visa delays caused by new regulations and procedures that were enacted to safeguard national security were having a negative impact on foreign students and scholars/researchers. The consequences of this visa policy include delays in research and increased costs of research. Other reported consequences include the increasing number of international conferences being held outside of the United States, and the fact that over one-third of available international students and academics are going to other countries. Because an estimated 70 to 80 percent of the visa delays related to science and engineering disciplines that are important to nanoscience and technology disciplines, continued delays would have a negative impact on the future development of this critical resource base. NAFSA has made a number of recommendations to address these issues.

A variety of institutions and laws mediate the flow of research and development in nanotechnology between global markets and users. The growth in nanotechnology-related commerce will depend on ongoing support for research and development programs that produce new products and services, supported by international laws that protect intellectual property rights, such as the Agreement on Trade-Related Aspects of Intellectual Property Rights (TRIPS), which was concluded in the 1994 Uruguay Round. TRIPS implementation continues to be slow as nations negotiate for their best interests. Less developed countries continue to seek ways to access advanced technologies and products without paying the monopolistic prices of those holding IP protection. This will be especially important if developments in nanotechnology make existing products and processes obsolete. The World Intellectual Property Organization (WIPO) promotes the use and protection of intellectual property and offers the principal forum for U.S., European Commission and Japan patent harmonization in key emerging science and technology areas.

The United States is party to international agreements because, on balance, they serve our national interests. They do, however, place some constraints and obligations on the government's policies for nanotechnology. TRIPS, for example, requires countries to tighten their laws on the protection of intellectual property rights. But it also contains language that requires countries to provide incentives to their companies to transfer technology to less-industrialized countries. Multinational companies may find it in their own interests to promote such transfers as part of a strategy of moving production capacity to lower-cost countries. To the extent that nanotechnology applications become commonplace in commerce, the technology will probably be transferred by private means, regardless of protectionist arguments. In nanotechnology, as in other areas of high technology, maintaining a competitive edge will require continued investment in new technologies that sustain economic development, since "older" technology will be embedded in global commodities that will be controlled by low-cost producers.

Action Recommendations

Investment in the National Nanotechnology Initiative must be sufficient, both in a broad range of research to advance the technology and in studies on the societal implications, so that the people of the world will gain the maximum benefit. Socio-legal research can contribute to resolving such issues as intellectual property rights at a policy level, but there are also practical issues such as whether legal shops should be created for the small businesses and minority-owned entrepreneurs who might otherwise be unable to benefit from the specialized expertise that large corporations can afford to hire. While companies are responsible for their actions and products, societal institutions must address broad environmental and health issues, as well. The widest possible cooperation will be required to address the specific ways in which existing international law, including human rights law, and security issues apply to nanotechnology developments.

- The advisory committee created by the 21st Century Nanotechnology Research and Development Act should perform a careful and rigorous analysis of the adequacy of current funding levels.

- The government should

  – significantly increase the funding available to understand the human health and environmental consequences of nanomaterials

  – review the adequacy of the current regulatory environment for nanomaterials, given the existence of size-dependent properties

- A conference should be organized to determine the role that nanotechnology could play in meeting the Millennium Development Goals adopted by the international community in 2000 [10], followed by a solicitation that would encourage joint research between U.S. and developing country researchers.

- An international forum should be established to allow discussion of IP, security and human rights in the public arena, including wide participation from NGOs and scholars from the global community, as well as industry and governments.

References

1. The Toxic Substances Control Act (TSCA) of 1976 (available at http://www.epa.gov/region5/defs/html/tsca.htm).

2. Supporting responsible nanotechnology research will benefit Europe's citizens, says head of unit, *CORDIS News* (June 24, 2003) (available at http://www.cordis.lu/nanotechnology/src/past-highlights.htm).

3. See page 23.

4. Constitution of the United States, Article I, Section 8, Clause 8.

5. R. P. Basler, *The Collected Works of Abraham Lincoln*, s.l., p.363: The Abraham Lincoln Association (1953) (available at http://www.hti.umich.edu/l/lincoln).

6. D. L. Burk, M. A. Lemley, Policy levers in patent law, *Virginia Law Review* **89**, 101-215 (2003).

7. M. C. Roco, Broader societal issues of nanotechnology, *J Nanoparticle Research* **5**, 181-189 (2003).

8. Z. Huang, H. Chen, A. Yip, G. Ng, F. Guo, Z. K. Chen, M. Roco, Longitudinal patent analysis for nanoscale science and engineering: Country, institution and technology field, *J Nanoparticle Research* **5**, 333-363 (2003).

9. J. R. Allison, M. A. Lemley, The growing complexity of the United States patent system, *Boston University Law Review* **82** (2002).

9. AUTM Licensing Survey: FY 2001 (summary available from http://www.autm.net/events/File/Surveys/01_Abridged_Survey.pdf).

10. United Nations Milenium Development Goals, http://www.un.org/millenniumgoals/R. R. Katz, P. Gold, *Justice Matters: Rescuing the Legal System for the Twenty-first Century*, Washington, D.C.: Discovery Institute (1997).

General References

S. E. Miller, The convergence of n: On nanotechnology, nanobiotechnology, and nanomedicine, *New York Law Journal* **230** (2003).

National Institutes of Health, *NIH Principles and Guidelines for Sharing of Biomedical Resources*, http://ott.od.nih.gov/res_tools.html.

R. Susskind, *The Future of Law: Facing the Challenges of Information Technology*, Oxford, UK: Oxford University Press (1996).

U.S. Government Accounting Office, *Technology Transfer: Administration of the Bayh-Dole Act by Research Universities*, Washington, D.C.: GAO (May 7, 1998).

Council on Governmental Relations, *The Bayh-Dole Act: A guide to the law and implementing regulations*, Washington, D.C.: Council on Governmental Relations (September 1999).

G. Brumfiel, As one door closes..., *Nature* **427**, 190-195 (2004).

THEME 9: INTERACTION WITH THE PUBLIC

Moderators: Davis Baird (University of South Carolina) and Catherine Alexander (National Nanotechnology Coordination Office)

Contributors: Carol Lynn Alpert, Elaine Bernard, David Berube, Toby Ten Eyck, Linda Goldenberg, Barbara Karn, Bruce Lewenstein, Jane Macoubrie, Cyrus Mody, Julia Moore, David Rejeski, Richard Smith, Christopher Toumey

Introduction

During the last 100 years, the impact of technology on our individual and social lives has been substantial. As a result, questions about the control and shaping of future developments in technology have become extremely important. There is widespread agreement in the scientific community and among policy makers that, given the importance of the technology, greater public involvement in technology-related decision making is appropriate, indeed necessary. As [former] NSF Director Rita Colwell so eloquently put it in her address to the Societal Implications Workshop, "We [as a nation] must design the future of our choice, not just of our making" [1].

Current State of Knowledge

Some social scientists and humanists have warned that the costs of not engaging the many stakeholders in nanotechnology's future are too great to ignore; furthermore, public engagement is imperative if nanotechnology is to fulfill its promise. Negative public attitudes toward nanotechnology could impede research and development, leaving the benefits of nanotechnology unrealized and the economic potential untapped or, worse, leaving the development of nanotechnology to countries and researchers who are not constrained by regulations and ethical norms held by most scientists worldwide.

Social scientists and humanists have also noted that radical technological change—and surely this includes nanotechnological change—drives radical social change, and with social change, there is often an unequal distribution of risks and of benefits, which can engender further opposition.

Finally, as with all technologies, nanotechnology has the potential to produce negative consequences. Research can anticipate some of these consequences, and self-control or government regulation can help avoid such anticipated negative consequences. But the possibility remains that not all consequences will be anticipated, and we need to be prepared for possible resistance to nanotechnology that could result from such unanticipated and unintended consequences. Involving the various concerned publics in dialogue about nanotechnology early in the process will mitigate resistance that might result from such unintended consequences.

What forms should public engagement take, however? While public dialogue serves the democratic ideal of the United States, policy makers and scientists are right to fear *uninformed* public input. Indeed, much of the public fears uninformed public input! Informed input, however, will promote the development of nanotechnology for the betterment of humankind and our environment.

Anticipated Developments

Informed input and the mechanisms to develop and encourage informed input are crucial for the beneficial development and application of nanotechnology. Unfortunately, the steps needed to achieve an informed population about technical matters such as nanotechnology are ill understood, and negative consequences can result from poorly conceived citizen participation. Thus while the risk of inaction is great, the risk of poorly conceived or executed action can be great as well.

As the following recommendations suggest, a major research focus must be how to best engage the public in dialogue on nanotechnology. Fundamental and applied research and ongoing evaluation will be necessary to effectively initiate and conduct such dialogue among stakeholders in nanotechnology's future.

The various components of public outreach and engagement discussed below are interdependent. For instance, various educational activities will have a clear role in public outreach. Research on educational materials, in turn, will influence educational activities. Further, as we learn more from our research about productive ways to engage the public, and indeed about current public attitudes toward nanotechnology, we will be better positioned to develop useful ways to bring stakeholders into dialogue. Conversely, that very dialogue will inform the research itself. Thus, each piece of the project outlined below interacts with each other piece. Coordination will be important to success of such public dialogue.

The National Nanotechnology Initiative can play an important role establishing itself as an honest broker in coordinating research and development in nanotechnology with public hopes and fears. The NNI needs to coordinate this work with related work at various centers such as the National Nanotechnology Infrastructure Network anchored at Cornell and Stanford and the soon-to-be-established Center for Nanotechnology in Society. Done well, various efforts will build informed and engaged public dialogue aimed at aligning the social, technological, and economic goals driving the development of nanotechnology. It will promote a smooth technological transition to the various nanotechnology applications that have the potential to dominate the 21st century. Moreover, the actual technologies that are developed will have public confidence as they are introduced in this ongoing process of technological change.

This is an ambitious project. It requires a significant research dimension along with a significant dimension of action, bringing the various stakeholders into fruitful dialogue. It requires careful attention to content and its presentation, and it requires careful attention to the ways that nanotechnology is covered in our educational system. This work will not be inexpensive.

It is imperative for the National Nanotechnology Initiative to play a central role, and to this end, for the NNI to invest significant resources in the project. However, additional funding partners, including those in the commercial sector, can and should be involved. That being said, care must be taken to involve the private sector in such a way as not to undermine the public's trust. The project cannot be seen simply as the public relations wing of nanotechnology-related business (or research).

Research and Evaluation Methodologies

Little appears to have been done toward establishing best practices regarding communication with the public about emerging technologies. A concerted effort should be undertaken to compile such best practices; there is much to learn about engaging the public in areas such as nanotechnology.

To engage effectively the various elements of the diverse population of the United States in dialogue about nanotechnology, we need to learn about what currently is known or believed about nanotechnology and how these beliefs are differentially distributed. We also need to learn the dimensions of public concern about nanotechnology, as well as the dimensions of excitement when these various publics imagine future nanotechnology applications. We need to assess the effectiveness of various engagement approaches and to develop new, effective approaches to public dialogue. In short, we need to learn much, we need to work hard to engage the various publics in dialogue about nanotechnology, and we need to evaluate the approaches as we do so.

While we have much to learn, we do not start from absolute zero. Some approaches to engaging various publics have been evaluated, and we have some idea of what the key issues in nanotechnology are likely to be. As work proceeds, information regarding key issues will be refined and will inform best practices for ways to engage the public on nanotechnology.

A variety of methods for understanding public knowledge and attitudes toward nanotechnology, *and* the impact of public opinion on nanotechnological development will be necessary. To begin, we need to develop survey data about awareness, understanding, and attitudes toward nanotechnology. Data from a variety of groups will be needed, running the gamut from those intensively engaged in developing the science and technology at the nanoscale to the many different segments of the public. Obtaining such information will be essential to pursuing all other initiatives outlined in this report.

More interactive and formalized ways of engaging the public and assessing attitudes include focus groups and events along the lines of the Danish consensus conferences [2]. These kinds of events have the advantage of both allowing for more intensive interaction between people over topics of concern and more methodological control to provide for assessment and refinement. A few such exercises in engaging the public have been tried in the United States, notably by the Loka Institute and a group of researchers at North Carolina State University [3]. Substantially more work of this kind must be built on these initial efforts.

Knowledge and concerns about nanotechnology are likely to be varied among the public. Consequently, it is vital that the educational materials and content developed in this project include the full spectrum of information, from the fascinating science and its promising applications to the possible consequences for our environment, our social structures, our bodies and daily life. Accordingly, content should be developed around these questions:

1. What is nanotechnology?

2. What are the potential nanotechnology applications?

3. How will nanotechnology impact the environment?

4. How will nanotechnology impact human society?

5. How will nanotechnology impact human bodies?

Within each of these main headings fall many sub-questions, the answers to which will form the basic content for our initial public dialogue on nanotechnology. As our research and our feedback from the public tell us more, we will revisit and revise.

Early efforts must also be directed to advance our understanding of nanotechnological risk. Here three dimensions of the state of the art concerning our understanding of risk are essential:

1. assessing nanotechnology-related risk

2. communicating risk

3. ascertaining risk perception

Communication of risk necessitates, of course, an assessment of such risks, including those associated with nanotechnology generally, but more specifically the risks attached to the creation, use, handling, and disposal of ultrafine particles. Social scientists and humanists also must examine how to frame risk communication regarding nanotechnology and determine how the various public responses will be influenced by the way the risks of nanotechnology are framed. It must be noted that, absent an assessment of the risks involved, communication on the subject could cause more harm than help. What is unknown and speculative can be more frightening than what is known.

Genuine risks must be dealt with in an expeditious, open, and honest manner. Equally imperative, this panel believes, the various segments of the broad public must be given a voice to their hopes and fears, and this voice must play a genuine role in the way that nanotechnology develops.

Information about nanotechnology must be distributed through a wide spectrum of avenues designed to reach as broad an audience as possible. These can be sorted into three groups:

1. passive mass media products

2. interactive public group initiatives

3. education initiatives

Research, Education, and Infrastructure Development

The mass media will play a role in conveying information about nanotechnology to the public through news organizations, broadcast programs, books and other entertainment media. Valuable mass media products for educational purposes would include, among other things, a television documentary on nanotechnology and press packages designed to help journalists articulate a more nuanced picture of nanotechnology. These products should aim at three goals:

1. to accurately convey what nanotechnology has to offer

2. to describe what the benefits and risks of nanotechnology could be

3. to describe the benefits and risks in a way that moves away from a simple "good versus bad" depiction of this technology

In addition to using the media, we also must understand the media. How do the standard media influence public attitudes toward nanotechnology? Research needs to be done on the reach of, and audience response to, various media products that concern nanotechnology. These products run the gamut from books, movies, television, and the Internet to newspapers and magazines. They can be framed as entertainment or news or something in between.

Given that standard media outlets—newspapers, radio, and television—are located between sources—advertisers and audiences—it is important to understand how these organizations choose and present stories. Methods of content analysis have been used to study stories that appear in the media, but this approach offers little insight into the processes by which the stories are constructed. Interviews with those in the news-making process can shed light on these processes. Further research is necessary, in particular, to learn which sources of nanotechnology information are considered most reliable and important to those who report the news.

Informal venues that educate and provide an open environment for exchanging information and views could draw people into thinking about, and expressing their thoughts on, nanotechnology. A multiplicity of venues for such interaction exist and should be utilized. These include "open houses" at nanoscience research centers, events at museums, discussions within civic or faith communities, and discussions at professional group meetings. Ultimately, such events can serve as seeds for further "discussions around the office coffee pot."

While these informal events are difficult to monitor, and would for this reason be difficult to formally assess, some efforts need to be made in developing means to do this.

Materials must be readily available to prepare discussion leaders at such events for conversations about the particular nanotechnology and its potential societal effects. To this end, materials on the web—e.g., downloadable presentations—should be made available for use at such events. The creation of a central resource of educational materials designed for a variety of levels will be invaluable to everyone engaged in public outreach. They would include briefing packets for the media, downloadable publications for the general public, and materials crafted appropriately for different school grades (K-20) to allow teachers to introduce nanotechnology—and in particular nanotechnology's societal implications—into their curricula.

It might be worth establishing an expertise bureau for societal issues related to nanotechnology, and holding media forums for journalists and the public. Interactive public group initiatives could include town hall style debates, non-credit extension courses on nanotechnology and its social and ethical consequences, museum events, civic or faith community events, nanocenter "open houses," and even nanotechnology-related movie events, for example, on the release of the movie version of Michael Crighton's novel, *Prey*.

More scientists and engineers should be trained to engage in public dialogue about nanotechnology. It is essential for those who are actively engaged in research at the nanoscale to be more actively engaged in public outreach on nanotechnology and to be conversant in the societal issues related to nanotechnology.

While theoretical studies of public engagement with—and influence on— technological development are important and will be useful, the work that will be most important will simultaneously teach researchers—including scientists and engineers—about public involvement with nanotechnology *and* inform the public about nanotechnology. Such applied studies will help to build reliable theory and will have important practical results.

The National Nanotechnology Initiative—working through the Nanoscale Science, Engineering, and Technology (NSET) Subcommittee of the National Science and Technology Council—should take the lead in advancing and coordinating work on the recommendations of this report. It is important that significant resources from the NNI be applied to these activities. More resources than are currently being allocated to this dimension of work on nanotechnology will be essential to creating fruitful dialogue. The NNI's website will become an essential resource in articulating and coordinating efforts toward establishing public dialogue on nanotechnology.

Beyond this, however, other partners must be enrolled in this work. This is necessary both because the NNI cannot carry the entire financial burden related to public outreach research and activities and because one of the advantages to enrolling other partners is their ability to help develop additional avenues for advancing public discussion on nanotechnology. These partners could include corporations, industry groups, professional associations, and regional economic development groups.

While a heterogeneous mix of partners is most advantageous to public engagement, contributions should be structured in such a way as to avoid any sense of outright advocacy. Central to the promotion of public dialogue on nanotechnology is the gaining of public trust that the information and the avenues for its exchange are open to all legitimate viewpoints, and not only those supporting a certain kind of development.

Action Recommendations

Publics need to be actively engaged in learning about nanotechnology. While all venues will be important, passive forms of education and outreach will be less successful in significantly raising awareness than active models. Whether passive media-centered approaches or active interaction-based deliberative approaches are used, this capacity will develop and use repositories of knowledge accessible to citizens, whereby one can obtain answers to specific questions about nanotechnology. This capacity will also include physical places, such as museums and science and technology centers, where citizens can meet and learn from each other's perspectives, including the perspectives of those pursuing nanotechnological research and of those engaged in developing policy for nanotechnology.

- A Federally sponsored workshop could serve to build a community around the issues and agenda concerning public outreach and provide a forum for the exchange of information between those primarily engaged in research into methodologies for engaging the public and those actually engaging the public.

- Research should be supported to develop various models of public involvement and interaction, to establish best practices for educating, communicating, and engaging diverse publics about nanotechnology.

- The NNI should embrace the goal of building capacity for public dialogue and make this goal central to the NNI's efforts to develop nanotechnology.

References

1. See page 11.

2. I.E. Andersen and B. Jæger, Danish participatory models, *Science and Public Policy* **26**, 331(1999).

3. See http://www.ncsu.edu/chass/communication/ciss/sponsored.html.

THEME 10: EDUCATION AND HUMAN RESOURCE DEVELOPMENT

Moderators: Michael Gorman (University of Virginia) and William Frascella (National Science Foundation)

Contributors: James Batterson, George Borjas, Tanwin Chang, Daniel Goroff, Judith Klein-Seetharaman, Kristen Kulinowski, Sharon Levin, Philip Lippel, John T. Neer, Deb Newberry, Ron Oaxaca, Paul Petersen, John Sargent, Bruce Seely, Paula Stephan, Sarah Turner

Introduction

Nanotechnology creates an opportunity to integrate education across physical science, technology, social sciences, and humanities. It is emblematic of new ways of thinking about the future and the workforce. Nanotechnology will enhance disciplinary depth and can help students at multiple levels—starting at the elementary level—see the fundamental connections among disciplines. For example, a nanotechnology revolution will be social as well as scientific and technological.

Therefore, it cuts across C. P. Snow's two cultures, the sciences and the humanities [1]. Students at all levels need to understand the coupling of society and technology. Students will be motivated to solve problems that combine the social and the technical, for example, realizing the potential for new environmental technologies. Nanotechnology studies, especially through the convergence of many fields of science and engineering at the nanoscale, will contribute to the mission of liberal education—to make students into critical thinkers, capable of participating in intelligent debates about how societies ought to be transformed.

Current State of Knowledge

One barrier to such liberal education is the way in which standards of education segment the K-12 curriculum, requiring teachers to prepare students for separate exams in different sciences and mathematics and to meet various standards, including those of the "No Child Left Behind" program. A short-term strategy involves using nanotechnology as means for facilitating learning within the framework of the standards to achieve coherence. One program that works within this model is offered by the Center for Biological and Environmental Nanotechnology (CBEN), an NSF Nanoscale Science and Engineering Center at Rice University. CBEN's teacher-training program includes a semester-long course where teachers learn how nanotechnology research projects rely upon the same concepts teachers present in their classrooms. Or, as CBEN Executive Director Dr. Kristen Kulinowski put it, "At its heart, nanotechnology is really just biology, chemistry, and physics." By illustrating that cutting-edge research is based upon the same science concepts within the standard science curricula, CBEN aims to reduce the barrier to introduction of nanotechnology concepts into the classroom. In other words, teachers won't have to teach new concepts, just use nanotechnology examples when they teach the familiar state curricula.

Another problem is the high rate at which students switch out of physical science and engineering majors and careers. Only 54 percent of originally enrolled men and 21 percent of originally enrolled women graduate with engineering degrees. Top reasons for switching out of engineering are poor teaching (98 percent), poor understanding of engineering careers (94 percent) and inadequate advising (81 percent), while curriculum overload and loss of interest in engineering are lesser components [2].

A longer-term solution is to transform standards without sacrificing competencies. In other words, students still need to learn skills such as laboratory experimentation, mathematics, and writing. They will also need to gain disciplinary knowledge, but not acquire fragmented disciplinary perspectives and biases. Some aspects of the current curriculum will have to be de-emphasized or eliminated. Problem-based learning, which can be more motivating for students, should be more widely used; it will help them understand the potential for careers in science and engineering.

Nanotechnology's influence on the curriculum should be significant, contributing to informed, educated publics emerging from our high schools and colleges, able to shape the direction of nanotechnology in beneficial ways.

Anticipated Developments

One aspect of the educational revolution will be the creation of a new generation of *interactional experts*. This term is used by Collins and Evans to describe the kind of expertise required to facilitate interaction among disciplines [3]. The convergence across multiple technologies makes this kind of expertise essential—especially since the convergence must include societal and ethical dimensions [4]. Interactional expertise complements, but does not replace, the need for disciplinary depth.

The *trading zone* serves as a good metaphor for interdisciplinary collaborations [5]. Consider, for example, the kinds of trading that went into proposals for the National Nanotechnology Infrastructure Network, where multiple institutions had to exchange resources and expertise in order to compete. Peter Galison has proposed that such trading zones require development of a creole, or reduced common language [6]. Effective exploration of societal dimensions of nanotechnology will require a new kind of researcher (whom we'll call a *nanocajun*) who explores societal dimensions across multiple disciplines and cultures.

Deep disciplinary knowledge plus a robust nanocajun create the necessary conditions for trading knowledge, resources, and policy recommendations across societal dimensions of nanotechnology. Effective trading zones depend not only on creoles, but also on interactional experts, who can act as translators and agents facilitating trade.

Therefore, disciplinary education is going to have to be complemented by training in how to work in interdisciplinary teams, a skill emphasized by the Accreditation Board for Engineering and Technology (ABET) [7] and also in the Science and Engineering Centers created by the NSF.

Developments in nanotechnology will transform industrial trading zones in ways that are hard to predict. Therefore, employees will have to be prepared to adapt. ABET guidelines, for example, emphasize the importance of continuous learning beyond school. From the industry perspective professional versatility and communication skills are among the most important characteristics sought.

New technologies often have disruptive effects on the workplace, by rendering some existing jobs obsolete while opening up new opportunities. Consider, for example, the fate of skilled telegraph operators when the telephone came into use, or the fate of vacuum tube manufacturers at the time of the introduction of the transistor. Nanotechnology, like information technology, has the potential to increase productivity in multiple areas. These increases will change the nature of work in many areas—as have productivity advances achieved through information technology.

Research and Evaluation Methodologies

Can nanotechnology education leverage integration across standards of learning, thereby creating more convergence among disciplines? Research needs to focus on the viability and transferability of integration strategies. What materials work best for which developmental and curricular levels? All attempts to create new educational systems need careful evaluation.

What kind of nanotechnology experts does industry need? What kind of job descriptions call for this expertise? Does a core body of knowledge exist that would prepare individuals to enter multiple jobs? Research needs to focus on the research versus the industrial versus the service workforces, keeping in mind that the boundaries between these areas are often blurred. What kinds of continuous learning are called for in each of these areas? What changing skills are actually needed at multiple career stages, including "soft" skills like working in multidisciplinary teams?

What models are there from other interdisciplinary areas? Remember the importance of labeling a new area of expertise and of providing evidence of certification.

Forecasting labor trends in technical fields is problematic, and this approach should not be recommended. We need to collect good data on the jobs that students are actually getting, and conduct a holistic, systems-level analysis of the skills and knowledge that are needed for the workforce. But young people selecting careers must get their own first-hand experience, rather than rely upon aggregate statistics assembled by experts. For example, NSF's Research Experiences for Undergraduates program involves college students in cutting-edge research, and this could include studies of the societal impacts of nanotechnology.

Research and development efforts to shape and apply the best methods for communicating nanoscale research and its potential benefits and related public policy and ethics issues should be encouraged. Public audiences could be engaged on an ongoing basis through high quality television, radio, and multimedia. Science museum partnerships with nanoscale research centers could bring public audiences and school groups in contact with future technologies and careers through guest researcher talks (see Figure 3.4), forums, and exhibits addressing challenges at the

Figure 3.4. Harvard physicist Charlie Marcus engages a live audience on quantum computing at the Museum of Science, Boston (courtesy of Museum of Science).

frontiers of research. The impact of these programs and the quality of the information disseminated should be systematically evaluated.

This kind of information should be directed toward user queries. There are many publics, not just one public, and information needs to be presented to stakeholder groups in a form that these groups can understand. This will encourage the formation of creoles for public engagement. For example, is information desired about a specific technological area, such as nanoelectronics, nanobio, or nanomaterials, or is information required about specific application areas, such as computation, medicine, or environmental impact, or is information desired about the economic aspects or public policy? In engaging publics, demographic information about the engaged would be desirable, such as age, education level, and major fields of expertise. It also would be useful to know why the information is sought. Such informational systems are, of course, very difficult to construct and would need to evolve over the years.

Research, Education, and Infrastructure Development

Because technology is changing at an ever-increasing pace, it is essential that students of all ages learn more than just basic skills and knowledge. They must also learn how to learn as a way of continuously adapting to a changing world. In elementary and secondary education, for example, this could involve greater participation in the learning process using problem-based challenges ("design a better X"), science fairs, role-playing games and model simulations. For the adult learner, regardless of the level of formal education already obtained, a variety of "continuing education" opportunities need to be made available by educational institutions, industries, and labor organizations.

Worker transition programs should be framed as opportunities to get in on the ground floor of a growing field. These programs should include working with different equipment to ensure transfer of tacit as well as explicit knowledge [8].

There also should be a focus on learning the emerging creoles that develop in these new areas.

The postdoctoral position has a long tradition in the United States. For example, for more than 30 years the typical career path of a research life scientist in the United States has involved obtaining a postdoctoral position upon receipt of the Ph.D. For new Ph.D.s with an interest in pursuing an academic career, the postdoctoral position has been an absolute necessity given that departments, when making tenure-track hires at the rank of assistant professor, direct their searches to the postdoctoral pool, not to those who have just received their degree.

Two dimensions of postdoctoral training have changed in recent years, however, leading to a dramatic increase in the number of postdoctoral fellows. One involves the increasing number of new Ph.D.s taking a first postdoctoral position, including individuals who have received their doctoral training abroad. The other involves a lengthening of the duration of the postdoctoral position. In early years, the postdoctoral position typically lasted only two years. This is no longer the case. For example, 35 percent of life science Ph.D.s observed in 1999 were in postdoctoral positions three to four years after graduation, compared to 12 percent in 1977; 20 percent held postdoctoral positions five to six years later, compared to 5 percent in 1977.

Recognition of the importance of the societal dimensions of nanotechnology creates an opportunity for training postdoctoral fellows in areas of technology that will be in demand from society. These postdoctoral fellows might be able to move more rapidly into leadership positions in policy, industry or academia.

Nanotechnology learning modules could be designed to be as compelling as a computer game. This idea goes back to George Leonard's *Education as Ecstasy* [9], which envisioned total immersion learning environments that are a step beyond current IT, but within range. Think of the way in which many adolescents interact in game universes, such as *Warcraft*. Why couldn't similarly compelling educational environments be constructed? Consider, for example, games like *Civilization, SimCity,* and *Europa Universalis.* Each of these allows participants to create and share mini-universes in which technology and society interact.

To turn games into educational tools, they need to be populated with activities that produce increased understanding. Consider, for example, a simulation that would provide an MRI map of a student's own body and would allow her or him to explore it.

Simulations could also combine face-to-face interaction with synchronous and asynchronous modes of communication over a network. A good example is virtual laboratories that connect to real ones, in which real experiments are simulated first on a computer and then students are given a chance to run them in a real laboratory, perhaps over long distance. So, for example, students could learn how to do simulated experiments on the effects of nanoparticles, and then be given the chance to participate in the design of an actual experiment that they could observe in real-time, if not in person.

Another example is provided by faculty and engineering students at the University of Virginia, who have designed an interactive simulation of the political and technological decisions involved in the space program. Students play specific roles in the simulation that have real-world analogies. For example, one group represents Congress, others represent private contractors like Boeing and Lockheed-Martin, still others represent NASA facilities like Jet Propulsion Lab and Goddard Space Flight Center, and one group represents a newspaper like *The Washington Post*. Students in the research facilities have to decide which technologies to pursue, fight for Federal funding, and contract launch capabilities from the companies. By the end, students have learned the way in which technology and politics are intertwined, and they have also learned how to function in multidisciplinary teams in which participants have allegiances to specific agencies and roles. Research in this simulation is done via a *Civilization*-type technology tree that makes certain advances prerequisites for others. A probabilistic model built into the software, which was built entirely by two undergraduates, determines the success of launches.

This crude simulation could be converted to a more sophisticated environment in which students explore the societal dimensions of nanotechnology. Students could be asked to form collaborative groups to compete for simulated funding for research centers and networks like the NSECs and the NNIN. The student groups would not have to be co-located; they could be brought together from across the country, or even from around the world. Budgets, research advances, and simulations of breakthroughs could be handled via software, but the meat of the simulation would be interpersonal interaction. The goal would be to get students from multiple disciplines and backgrounds to imagine possible futures for nanotechnology and hammer out the details of how these futures could be realized.

Action Recommendations

- New curricula and educational materials should be created that will move beyond current standards of learning to encourage convergence among disciplines and the development of interactional expertise.

- Government, industry, and academia should find ways to reduce and defray the cost of worker retraining, building on such existing institutions as community colleges.

- Regional centers for societal and ethical implications of nanotechnology should be established, to serve as test beds for long-term nanotechnology curriculum integration where students and teachers could work with scientists.

References

1. C. P. Snow, *The Two Cultures and the Scientific Revolution*, Cambridge, UK: Cambridge University Press (1993).

2. B. M. Kramer, Time to rethink engineering education? Speech presented at the University of South Florida, March 6 (2003).

3. H. M. Collins, R. Evans, The third wave of science studies, *Social Studies of Science* **32**(2), 235-296 (2002).

4. M. E. Gorman, Combining the social and the nano: A model for converging technologies, in M. C. Roco, W. S. Bainbridge, eds., *Converging Technologies for Improving Human Performance: Nanotechnology, Biotechnology, Information Technology and Cognitive Science*, Dordrecht: Springer (formerly Kluwer) (2003).

5. M. E. Gorman, Expanding the trading zones for convergent technologies, in M. C. Roco, W. S. Bainbridge, eds., *Converging Technologies for Improving Human Performance: Nanotechnology, Biotechnology, Information Technology and Cognitive Science*, Dordrecht: Springer (formerly Kluwer) (2003).

6. P. Galison, *Image & Logic: A Material Culture of Microphysics*, Chicago: University of Chicago Press (1997).

7. M. E. Gorman, Turning students into professionals: Types of knowledge and ABET engineering criteria, *Journal of Engineering Education* **91**(3), 339-344 (2002).

8. M. E. Gorman, Types of knowledge and their roles in technology transfer, *The Journal of Technology Transfer* **27**(3), 219-231 (2002).

9. G. B. Leonard, *Education and Ecstasy*, New York: Delacorte Press (1968).

10. M. C. Roco, Nanotechnology: A frontier for engineering education, *Int J Engng Ed* **18**(5), 488 (2002).

Appendix A. Agenda

National Nanotechnology Initiative (NNI) Workshop on
Societal Implications of Nanoscience and Nanotechnology

National Science Foundation

4201 Wilson Blvd., Arlington, VA 22230

December 3-5, 2003

DAY 1 (DECEMBER 3, 2003), Plenary Session, Room 375

12:00 p.m.   Refreshments

1:00 p.m.   *Welcome,* Rita Colwell, NSF Director
Charge to the workshop, Mihail Roco, NSET, NSF

National Endeavor
Moderator: Mihail Roco

1:15 pm   *Nanotechnology: a national endeavor*
John Marburger, Director, Office of Science and Technology Policy

1:30 p.m.   *Technological and economic goals*
Phillip Bond, Undersecretary for Technology,
U.S. Dept. of Commerce

1:55 pm   *Science and education vision for nanoscience and nanotechnology*
George Whitesides, Harvard University

2:20 p.m.   Coffee

Technological and Societal Goals
Moderator: Rachelle Hollander

2:40 p.m.   *Industry implications of nanotechnology*
Tom Theis, IBM

3:00 p.m.   *Nanotechnology and society*
Roger Kasperson, Clark University, Stockholm Environment Institute

3:20 p.m.   *Social science approaches for assessing nanotechnology*
Lynn Zucker, UCLA

3:40 p.m.   Coffee

Broader Implications
Moderator: William Bainbridge

4:00 p.m.   *Nanotechnology Implications on quality of life: Medicine,*
            environmental, cognition, communication, and other areas
            Carlo Montemagno, UCLA

4:20 p.m.   *Ethical, philosophical issues*
                Vivian Weil, IIT

4:40 p.m.   *Navigating Nano through Society*
                Davis Baird, University of South Carolina

Evening

5:30 p.m.   Reception – The Front Page Grille in the NSF Building Atrium

6:30 p.m.   Group Dinner – Hilton Hotel Banquet Room
            (connected via skywalk to NSF)

DAY 2 (DECEMBER 4, 2003)

We will break into five separate (parallel) panels to explore future opportunities and potential breakthroughs in selected sub-fields. For this part of the program, participants are encouraged to come prepared with two pages and two slides (maximum) for a five-minute (maximum) presentation on their ideas for the future of the relevant field.

Plenary presentation, Room 375
Moderator: Mihail Roco, NSF, NSET

8:00 a.m.   *Technological convergence from the nanoscale (NBIC)*
                James Spohrer, IBM

Panels, A: Current Issues/Topics in Setting a Research Agenda

8:20 – 11:00 a.m.

Impact of nanotechnology on productivity and equity, Room 375

Moderators: Mihail Roco (NSET, NSF) and Marie Thursby (GIT)

Contributors: Evelyn Hu, Georg G. A. Böhm, George Thompson, Mark Andrews, Mark Modzelewski; John Belk, Gregory Tassey, Jeff Stanton, Brian Valentine, William Boulton, Ray Tsui, Louis Hornyak, Peter Hébert, James Canton, Jim Adams, Brad DeLong, Jared Bernstein, Sarah Turner, Richard Freeman, Larry Iannaccone, Robin Hanson

Nanotechnology implications on quality of life (medical, environmental, cognition, communication, etc.): Nanotechnology goals and unintended consequences, Room 340

Moderators: Carlo Montemagno (UCLA) and Michael Heller (UCSD)
Contributors: Steven Papermaster, David A. Diehl, Rosalyn Berne, Toby Ten Eyck, Barbara Karn, Kristen Kulinowski, Jeff Schloss, Hongda Chen, Donald Marlowe, Stan Brown, Sean Murdock, Dick Livingston, Richard Smith, Elaine Bernard, Tanwin Chang, Nila Bhakuni, Stephan Herrera, Günter Oberdörster

Ethical, historical, governance, philosophical implications, risk, and uncertainty, Room 370

Moderators: Vivian Weil (IIT) and Rachelle Hollander (NSF)
Contributors: M. Kathleen Behrens, Albert Teich, Eleanor Singer, Deb Newberry, Carol Lynn Alpert; Philip Sayre; Dan Jones, Sheila Jasanoff, Robert McGinn, Julia Moore, Jane Macoubrie, Frank Laird, Arthur Caplan, Daniel Goroff, John T. Trumpbour, Bruce Lewenstein

Converging technologies and their societal implications, Room 380

Moderators: John Sargent (DOC) and Lynne Zucker (UCLA)

Contributors: James R. von Ehr II, Judith Klein-Seetharaman, Ilesanmi Adesida, Sonia E. Miller, Roger Kasperson, David Rejeski, Sharon Levin, Paula Stephan, Cyrus Mody

National security, space exploration, Room 390

Moderators: Delores Etter (DOD) and Jim Murday (ONR)

Contributors: Kwan Kwok, James Batterson, Judith Reppy, John T. Neer, W.M. Tolles, Jim Murday, Scott McNeil, Minoo Dastoor, Martin Carr, Keith Ward, Cliff Lau, George Borjas, Ron Oaxaca, Grant Black

9:30 a.m.   Coffee Break

11:00 a.m.   Plenary presentations of summaries 1-5

12:30 p.m.   Working Lunch (lunch brought in the room)

DAY 2 (DECEMBER 4, 2003) continued
Panels, B:

1:30-4:00 p.m

Interaction with the public and social networks, Room 375
Moderators: David Baird (USC) and Cate Alexander (NNCO)
Contributors: Steven Papermaster, Albert Teich, Julia Moore, Toby Ten Eyck, Jane Macoubrie, Carol Lynn Alpert, Barbara Karn, Mark Modzelewski, Rosalyn Berne,

Dan Jones, David Berube, Bruce Lewenstein, David Rejeski, Elaine Bernard, Jared Bernstein, Cyrus Mody

Future economic scenarios, Room 340

Moderators: Gregory Tassey (NIST) and Michael Darby (UCLA)

Contributors: Mark Andrews, Robin Hanson, Ilesanmi Adesida, Scott McNeil, Georg G. A. Böhm, Judith Reppy, Brian Valentine, Hongda Chen, David Mowery, Sean Murdock, Linda Parker, Peter Hébert, Jim Adams, Brad DeLong, Richard Freeman, Louis Hornyak

Future social scenarios, Room 370

Moderators: Bill Bainbridge (NSF) and Roger Kasperson (Clark Univ.)

Contributors: Frank Laird, Rosalyn Berne, Jeff Schloss, John Belk, Jeff Stanton, John Miller, James Canton, Dick Livingston, Arthur Caplan, John T. Trumpbour, Stephan Herrera, Günter Oberdörster, Larry Iannaccone

Public policy, legal (patents, civic, etc.), and international aspects, Room 380

Moderators: Evelyn Hu (UCSB) and James Rudd (NSF)

Contributors: Sonia E. Miller, Sheila Jasanoff, Philip Sayre, George Thompson, James R. von Ehr II, V. Weil, Robert McGinn, W.M. Tolles, William Boulton, E. Jennings Taylor, Stan Brown, Ray Tsui, Richard Smith, Nila Bhakuni, Michael Heller

Education and human development, Room 390

Moderators: Michael Gorman (U. VA) and William Frascella (NSF)

Contributors: Paul Petersen, Bruce Seely, James Batterson, Deb Newberry, Kristen Kulinowski, Paula Stephan, Sharon Levin, Philip Lippel, Ron Oaxaca, George Borjas, Tanwin Chang, Daniel Goroff, Sarah Turner, Judith Klein-Seetharaman, John T. Neer, John Sargent

Plenary Presentations, Room 375

4:10 p.m.   *Economical trends and nanotechnology development*
           Brad DeLong, UC Berkeley

4:35 p.m.   *Human resources for nanotechnology*
           Paula Stephan, Georgia State Univ.

5:00 p.m.   Plenary presentations of summaries 6-10, Room 375

7:00 p.m.   Group dinner – The Front Page Restaurant & Grille
           in NSF Building Atrium

DAY 3 (DECEMBER 5, 2003) Plenary Session, Room 375

8.00 a.m.   Plenary discussion

- Definition of research and education challenges
- Recommendations for future R&D, infrastructure and education needs, societal preparation, etc.
- Plan for report preparation and agenda for the remainder of the day (M. Roco)

9:15 a.m.   Coffee break, proceed to breakout rooms

Breakout Sessions, Rooms 340, 370, 380, 390
9:30 a.m.   Dec. 4 panels meet individually to refine summaries presented at end of previous days' discussions and agree on report drafting assignments for report chapters that will arise from each panel session.

Plenary Session
11:00 a.m.   Each of the 10 panels present refined summary and outline/report writing assignments to the full group back in Rm. 375 (~5 minutes for each group). Plenary discussion to provide feedback, mid-course correction to these proposed outlines and assignments.

Optional Luncheon Session, Room 375
12:00 noon   Institutional Implications of Government Science Initiatives (15 minute prepared talks plus five minutes for questions/discussion after each talk)

Plenary Session, Room 375

12:20 p.m.   *Historical Comparisons for Anticipating Public Reactions to Nanotechnology*
Christopher Toumey, Univ. of South Carolina

12:40 p.m.   *Past Experiences*
Alex Roland, Duke Univ.

1:00 p.m.   *Present Adjustments*
Toby Smith, AAU

1:20 p.m.   *Future Perspectives: The Role of National Research Initiatives*
Tom Kalil, UC Berkeley

1:40 p.m.   *Congressional Perspective: Societal Implications Issues in the Nanotechnology Act*
David Goldston, Chief of Staff, House Science Committee

Breakout Sessions, Rooms 340, 370, 380, 390

2:00 p.m.   Re-group to breakout rooms for resumption of drafting sessions

Plenary Session, Room 375

3:45 p.m.   Concluding remarks (M. Roco)

4:00 p.m.   Adjourn

APPENDIX B. PARTICIPANTS AND CONTRIBUTORS*

James Adams
Rensselaer Polytechnic Institute

Ilesanmi Adesida
University of Illinois at Urbana-Champaign

Norris Alderson
Food and Drug Admimistration

Catherine Alexander
National Nanotechnology Coordination Office

Carol Lynn Alpert
Museum of Science, Boston

Mark Andrews
Caterpillar Inc.

William S. Bainbridge
National Science Foundation

Davis Baird
University of South Carolina

James Batterson
National Aeronautics and Space Administration

Matthew J. Bauer
National Science Foundation

M. Kathleen Behrens
Robertson Stephens & Co. and PCAST representative

John H. Belk
The Boeing Company

Robert Beoge
ASTRA

Elaine Bernard
Harvard University

Rosalyn W. Berne
University of Virginia

Jared Bernstein
Economic Policy Institute

David M. Berube
University of South Carolina

Nila Bhakuni
Harvard University

Grant Black
Georgia State University

Georg G. A. Böhm
Bridgestone / Firestone Research, LLC

George Borjas
Harvard University

William Boulton
Auburn University

Stanley A. Brown
Food and Drug Administration

James Canton
Institute for Global Futures

Arthur L. Caplan
University of Pennsylvania

Martin Carr
DCI

Tanwin Chang
National Bureau of Economic Research

Hongda Chen
U.S. Department of Agriculture

Ken Chung
Harvard University

---

* Institutional affiliations as of December 2003.

Julia Clark
National Science Foundation

Michael Darby
University of California, Los Angeles

Minoo N. Dastoor
National Aeronautics and Space
Administration

Michael E. Davey
Congressional Research Service

J. Bradford DeLong
University of California, Berkeley

David A. Diehl
PPG Industries, Inc.

David Donnelly
Quinnipiac University

Adam M. Eisgrau, Vice President
Flanagan Consulting LLC

Delores M. Etter
United States Naval Academy

Erik Fisher
University of Colorado at Boulder

William J. Frascella
National Science Foundation

Richard Freeman
Harvard University and National
Bureau of Economic Research

Linda Goldenberg
The University of Calgary

Michael E. Gorman
University of Virginia

Daniel Goroff
Harvard University

Robin D. Hanson
George Mason University

Peter Hébert
Lux Capital

Michael Heller
University of California, San Diego

Stephan Herrera
The Economist

Geoffrey M. Holdridge
National Nanotechnology Coordination
Office

G. Louis Hornyak
University of Denver

Evelyn L. Hu
University of California

Laurence Iannaccone
George Mason University

Sheila Jasanoff
Harvard University

Thomas A. Kalil
University of California, Berkeley

Sally M. Kane
National Science Foundation

Barbara Karn
Environmental Protection Agency

Roger Kasperson
Clark University and Stockholm
Environment Institute

Judith Klein-Seetharaman
University of Pittsburgh Medical
School and Carnegie Mellon University

Bruce Kramer
National Science Foundation

Kristen M. Kulinowski
Rice University

Frank N. Laird
University of Denver

Cliff Lau
U.S. Department of Defense

Appendix B. Participants and Contributors

Sharon Levin
University of Missouri

Bruce Lewenstein
Cornell University

Soo-Siang Lim
National Science Foundation

Philip H. Lippel
AAAS/National Science Foundation

Richard A. Livingston
Federal Highway Administration

Jane Macoubrie
North Carolina State University

Donald E. Marlowe
Food and Drug Administration

Robert McGinn
Stanford University

Scott McNeil
SAIC

Mark Menna
The Nanotechnology Policy Forum

John C. Miller
U.S. Department of Energy

Sonia E. Miller
Converging Technologies Bar
Association

Cyrus Mody
Cornell University

Mark Modzelewski
NanoBusiness Alliance

Carlo Montemagno
University of California, Los Angeles

Julia Moore
National Science Foundation

David C. Mowery
Univ. of California, Berkeley

Sean J. Murdock
AtomWorks

Kesh Narayanan
National Science Foundation

John T. Neer
Lockheed Martin Space Systems
Company

Deb Newberry
The NanoTechnology Group, Inc.

Ron Oaxaca
University of Arizona

Günter Oberdörster
University of Rochester

Steven Papermaster
Powershift Ventures and PCAST
representative

Linda Parker
National Science Foundation

Robert Pearson
University of Pennsylvania

Paul Petersen
Rochester Institute of Technology

Joseph Reed
National Science Foundation

David Rejeski
Woodrow Wilson Center for Scholars

Judith Reppy
Cornell University

Glenn Rhoades
Colorado Nanotechnology Initiative

Lindsey Rich
National Science Foundation

Mihail C. Roco
National Science Foundation

James Rudd
National Science Foundation

Cheryl L. Sabourin
General Electric

John Sargent
Department of Commerce

Nora Savage
Environmental Protection Agency

Philip Sayre
Environmental Protection Agency

Jeffery Schloss
National Institutes of Health

Bruce E. Seely
Michigan Technological University

Richard H. Smith
The Nanotechnology Policy Forum

Jeffrey M. Stanton
Syracuse University

Paula Stephan
Georgia State University

Barry Sullivan
International Engineering Consortium

E. Jennings Taylor
Faraday Technology, Inc.

Albert H. Teich
American Association for the
Advancement of Science

Toby A. Ten Eyck
Michigan State University

Thomas N. Theis
IBM

George Thompson
Intel Corporation

Marie C. Thursby
Georgia Institute of Technology

W. M. Tolles
Consultant

Christopher Toumey
University of South Carolina

John T. Trumpbour
Harvard University

Ray Tsui
Motorola Labs

Sarah Turner
University of Virginia

Mark Verbrugge
GM Research and Development

James von Ehr II
Zyvex Corporation

Keith B. Ward
Department of Homeland Security

Vivian Weil
Illinois Institute of Technology

Svetlana Zemskova
Caterpillar, Inc.

Lynn Zucker
University of California, Los Angeles

INDEX